BASIC ARITHMETIC

BASIC ARITHMETIC

by
Himmat Banik | Amitava Agarwal

Wordpen Academics
Publishing Science That Matter

ISBN: 978-93-6084-187-4 (Hardback)

Published by : **WORDPEN ACADEMICS**
4736/23, Ansari Road, Darya Ganj,
New Delhi – 110 002
Phone: +91-011-35004336
E-mail: info@wordpenacademics.com

Printed at **:** **Replika Press Pvt. Ltd.**

Exclusive Distributor: ***Bio-Green Books, New Delhi***

Preface

In the realm of mathematics, arithmetic serves as the cornerstone upon which all other mathematical concepts are built. From the simplest of counting exercises to the most complex calculations, arithmetic provides the essential tools and techniques for understanding numerical relationships and solving practical problems in everyday life. As we embark on the journey of learning arithmetic, we lay the foundation for a lifetime of mathematical exploration and problem-solving skills.

This book, titled "Basic Arithmetic," is designed as a comprehensive guide to mastering the fundamental principles of arithmetic. Whether you are a student seeking to build a solid mathematical foundation, an educator looking for effective teaching resources, or an adult learner eager to sharpen your numeracy skills, this book offers a structured and accessible approach to learning arithmetic concepts.

From the four basic operations of addition, subtraction, multiplication, and division to more advanced topics such as fractions, decimals, and percentages, "Basic Arithmetic" covers a wide range of essential mathematical concepts in a clear and concise manner. Through step-by-step explanations, worked examples, and interactive exercises, readers are guided through the principles and techniques of arithmetic, gaining confidence and proficiency with each new concept.

Moreover, this book emphasizes the real-world applications of arithmetic, demonstrating how mathematical principles are used in everyday scenarios such as budgeting, shopping, cooking, and problem-solving. By highlighting the practical relevance of arithmetic skills, we aim to motivate learners to engage with the material and appreciate the value of mathematics in their daily lives.

As we embark on this journey of mathematical discovery, it is our hope that "Basic Arithmetic" will serve as a valuable resource for learners of all ages and backgrounds. Whether you are a beginner exploring the fundamentals of arithmetic or seeking to reinforce your existing knowledge, this book invites you to embark on a journey of mathematical exploration, discovery, and mastery. Together, let us unlock the power of arithmetic and lay the groundwork for a lifetime of mathematical learning and problem-solving.

Himmat Banik | Amitava Agarwal

Contents

1

Introduction: Numbers

Numbers are an integral part of our everyday lives, right from the number of hours we sleep at night to the number of rounds we run around the racing track and much more. In math, numbers can be even and odd numbers, prime and composite numbers, decimals, fractions, rational and irrational numbers, natural numbers, integers, real numbers, rational numbers, irrational numbers, and whole numbers. In this chapter, we will get an introduction to the different types of numbers and to all the concepts related to them.

A number is a basic component of mathematics. Numbers are used for counting, measuring, keeping things in order, indexing, etc. We have different types of numbers based on their properties such as natural numbers, whole numbers, rational and irrational numbers, integers, real numbers, complex numbers, even and odd numbers, etc. We can apply the basic fundamental arithmetic operations of numbers and determine the resulting number. Initially, tally marks were used before the use of numbers. Let us now introduce the concept of numbers and understand their different types and their properties.

Numbers form the basis of mathematics. We should befriend numbers in order to understand math. Numbers are of various kinds. We have a long list that includes ordinal numbers, consecutive numbers, odd numbers, even numbers, natural numbers, whole numbers, integers, real numbers, rational numbers, irrational numbers, and complex numbers.

Along with numbers, we come across the interesting world of factors and multiples. This world includes prime numbers, composite numbers, co-prime numbers, perfect numbers HCF, LCM, and prime factorization.

A number is a mathematical object used to count, measure, and label. The original examples are the natural numbers 1, 2, 3, 4, and so forth. Numbers can be represented in language with number words. More universally, individual numbers can be represented by symbols, called numerals; for example, "5" is a numeral that represents the number five. As only a relatively small number of symbols can be memorized, basic numerals are commonly organized in a numeral system, which is an organized way to represent any number. The most common numeral system is the Hindu–Arabic numeral system, which allows for the representation of any number using a combination of ten fundamental numeric symbols, called digits. In addition to their use in counting and measuring, numerals are often used for labels (as with telephone numbers), for ordering (as with serial numbers), and for codes (as with ISBNs). In common usage, a numeral is not clearly distinguished from the number that it represents.

In mathematics, the notion of a number has been extended over the centuries to include 0, negative numbers, rational numbers such as one half $\left(\frac{1}{2}\right)$, real numbers such as the square root of ($\sqrt{2}$) and π, and complex numbers which extend the real numbers with a square root of "1 (and its combinations with real numbers by adding or subtracting its multiples). Calculations with numbers are done with arithmetical operations, the most familiar being addition, subtraction, multiplication, division, and exponentiation. Their study or usage is called arithmetic, a term which may also refer to number theory, the study of the properties of numbers.

Besides their practical uses, numbers have cultural significance throughout the world. For example, in Western society, the number 13 is often regarded as unlucky, and "a million" may signify "a lot" rather than an exact quantity. Though it is now regarded as pseudoscience, belief in a mystical significance of numbers, known as numerology, permeated ancient and medieval thought. Numerology heavily influenced the development of Greek mathematics, stimulating the investigation of many problems in number theory which are still of interest today.

During the 19th century, mathematicians began to develop many different abstractions which share certain properties of numbers, and may be seen as extending the concept. Among the first were the hypercomplex numbers, which consist of various extensions or modifications of the complex number system. In modern mathematics, number systems are considered important special examples of more general algebraic structures such as rings and fields, and the application of the term "number" is a matter of convention, without fundamental significance.

PRE-NUMBER MATH

Building pre-number math skills is a prerequisite to understanding numbers. Pre-number skills like matching, sorting, classifying, ordering, and comparing sets the stage to build a strong number sense. Pre-number Math skills are built-in preschool years. Kids learn how to stand before they start taking small steps. In the same way, the pre-number concept is very important for them to start understanding Mathematics. In this section, we will cover the different pre-number concepts like Matching and Sorting, Comparing and Ordering, Classification, and Shapes and patterns.

Example: Observe the figure given below which shows 2 columns. The left column displays the numbers 1 to 4. The right column displays rows of items. The numbers are matched to the quantities they represent. This is an essential skill for children aged 3 to 4 years.

NUMBER NAMES

Number names are used to represent numbers in an alphabetical format. A specific word is used to refer to each number. To write a number in words in English, we should know the place value of each digit in the number.

Example: Observe the figure given below which shows that 23 is written as 'twenty-three' in the number names form.

History of Numbers

Numerals

Numbers should be distinguished from numerals, the symbols used to represent numbers. The Egyptians invented the first ciphered numeral system, and the Greeks followed by mapping their counting numbers onto Ionian and Doric alphabets. Roman numerals, a system that used combinations of letters from the Roman alphabet, remained dominant in Europe until the spread of the superior Hindu–Arabic numeral system around the late 14th century, and the Hindu–Arabic numeral system remains the most common system for representing numbers in the world today. The key to the effectiveness of the system was the symbol for zero, which was developed by ancient Indian mathematicians around 500 AD.

First use of numbers

Bones and other artifacts have been discovered with marks cut into them that many believe are tally marks. These tally marks may have been used for counting elapsed time, such as numbers of days, lunar cycles or keeping records of quantities, such as of animals.

A tallying system has no concept of place value (as in modern decimal notation), which limits its representation of large numbers. Nonetheless tallying systems are considered the first kind of abstract numeral system.

The first known system with place value was the Mesopotamian base 60 system (c. 3400 BC) and the earliest known base 10 system dates to 3100 BC in Egypt.

Zero

The first known documented use of zero dates to AD 628, and appeared in the Brâhmasphumasiddhânta, the main work of the Indian mathematician Brahmagupta. He treated 0 as a number and discussed operations involving it, including division. By this time (the 7th century) the concept had clearly reached Cambodia as Khmer numerals, and documentation shows the idea later spreading to China and the Islamic world.

Brahmagupta's Brâhmasphumasiddhânta is the first book that mentions zero as a number, hence Brahmagupta is usually considered the first to formulate the concept of zero. He gave rules of using zero with negative and positive numbers, such as "zero plus a positive number is a positive number, and a negative number plus zero is the negative number." The Brâhmasphumasiddhânta is the earliest known text to treat zero as a number in its own right, rather than as simply a placeholder digit in representing another number as was done by the Babylonians or as a symbol for a lack of quantity as was done by Ptolemy and the Romans.

The use of 0 as a number should be distinguished from its use as a placeholder numeral in place-value systems. Many ancient texts used 0. Babylonian and Egyptian texts used it. Egyptians used the word nfr to denote zero balance in double entry accounting. Indian texts used a Sanskrit word Shunye or shunya to refer to the concept of void. In mathematics texts this word often refers to the number zero. In a similar vein, PâGini (5th century BC) used the null (zero) operator in the Ashtadhyayi, an early example of an algebraic grammar for the Sanskrit language (also see Pingala).

There are other uses of zero before Brahmagupta, though the documentation is not as complete as it is in the Brâhmasphumasiddhânta.

Records show that the Ancient Greeks seemed unsure about the status of 0 as a number: they asked themselves "how can 'nothing' be something?" leading to interesting philosophical and, by the Medieval period, religious arguments about the nature and existence of 0 and the vacuum. The paradoxes of Zeno of Elea depend in part on the uncertain interpretation of 0. (The ancient Greeks even questioned whether 1 was a number.)

The late Olmec people of south-central Mexico began to use a symbol for zero, a shell glyph, in the New World, possibly by the 4th century BC but certainly by 40 BC, which became an integral part of Maya numerals and the Maya calendar. Maya arithmetic used base 4 and base 5 written as base 20. George I. Sánchez in 1961 reported a base 4, base 5 "finger" abacus.

By 130 AD, Ptolemy, influenced by Hipparchus and the Babylonians, was using a symbol for 0 (a small circle with a long overbar) within a sexagesimal numeral system otherwise using alphabetic Greek numerals. Because it was used alone, not as just a placeholder, this Hellenistic zero was the first documented use of a true zero in the Old World. In later Byzantine manuscripts of his Syntaxis Mathematica (Almagest), the Hellenistic zero had morphed into the Greek letter Omicron (otherwise meaning 70).

Another true zero was used in tables alongside Roman numerals by 525 (first known use by Dionysius Exiguus), but as a word, nulla meaning nothing, not as a symbol. When division produced 0 as a remainder, nihil, also meaning nothing, was used. These medieval zeros were used by all future medieval computists (calculators of Easter). An isolated use of their initial, N, was used in a table of Roman numerals by Bede or a colleague about 725, a true zero symbol.

NEGATIVE NUMBERS

The abstract concept of negative numbers was recognized as early as 100–50 BC in China. The Nine Chapters on the Mathematical Art contains methods for finding the areas of figures; red rods were used to denote positive coefficients, black for negative. The first reference in a Western work was in the 3rd century AD in Greece. Diophantus referred to the equation equivalent to $4x + 20 = 0$ (the solution is negative) in Arithmetica, saying that the equation gave an absurd result.

During the 600s, negative numbers were in use in India to represent debts. Diophantus' previous reference was discussed more explicitly by Indian mathematician Brahmagupta, in Brâhmasphumasiddhânta in 628, who used negative numbers to produce the general form quadratic formula that remains in use today. However, in the 12th century in India, Bhaskara gives negative roots for quadratic equations but says the negative value "is in this case not to be taken, for it is inadequate; people do not approve of negative roots".

European mathematicians, for the most part, resisted the concept of negative numbers until the 17th century, although Fibonacci allowed negative solutions in financial problems where they could be interpreted as debts (chapter 13 of Liber Abaci, 1202) and later as losses (in Flos). René Descartes called them false roots as they cropped up in algebraic polynomials yet he found a way to swap true roots and false roots as well. At the same time, the Chinese were indicating negative numbers by drawing a diagonal stroke through the right-most non-zero digit of the corresponding positive number's numeral. The first use of negative numbers in a European work was by Nicolas Chuquet during the 15th century. He used them as exponents, but referred to them as "absurd numbers".

As recently as the 18th century, it was common practice to ignore any negative results returned by equations on the assumption that they were meaningless.

Rational numbers

It is likely that the concept of fractional numbers dates to prehistoric times. The Ancient Egyptians used their Egyptian fraction notation for rational numbers in mathematical texts such as the Rhind Mathematical Papyrus and the Kahun Papyrus. Classical Greek and Indian mathematicians made studies of the theory of rational numbers, as part of the general study of number theory. The best known of these is Euclid's Elements, dating to roughly 300 BC. Of the Indian texts, the most relevant is the Sthananga Sutra, which also covers number theory as part of a general study of mathematics.

The concept of decimal fractions is closely linked with decimal place-value notation; the two seem to have developed in tandem. For example, it is common for the Jain math sutra to include calculations of decimal-fraction approximations to pi or the square root of 2. Similarly, Babylonian math texts used sexagesimal (base 60) fractions with great frequency.

Irrational numbers

The earliest known use of irrational numbers was in the Indian Sulba Sutras composed between 800 and 500 BC. The first existence proofs of irrational numbers is usually attributed to Pythagoras, more specifically to the Pythagorean Hippasus of Metapontum, who produced a (most likely geometrical) proof of the irrationality of the square root of 2. The story goes that Hippasus discovered irrational numbers when trying to represent the square root of 2 as a fraction. However, Pythagoras believed in the absoluteness of numbers, and could not accept the existence of irrational numbers. He could not disprove their existence through logic, but he could not accept irrational numbers, and so, allegedly and frequently reported, he sentenced Hippasus to death by drowning, to impede spreading of this disconcerting news.

The 16th century brought final European acceptance of negative integral and fractional numbers. By the 17th century, mathematicians generally used decimal fractions with modern notation. It was not, however, until the 19th century that mathematicians separated irrationals into algebraic and transcendental parts, and once more undertook the scientific study of irrationals. It had remained almost dormant since Euclid. In 1872, the publication of the theories of Karl Weierstrass (by his pupil E. Kossak), Eduard Heine, Georg Cantor, and Richard Dedekind was brought about. In 1869, Charles Méray had taken the same point of departure as Heine, but the theory is generally referred to the year 1872. Weierstrass's method was completely set forth by Salvatore Pincherle (1880), and Dedekind's has received additional prominence through the author's later work (1888) and endorsement by Paul Tannery (1894). Weierstrass, Cantor, and Heine base their theories on infinite series, while Dedekind founds his on the idea of a cut (Schnitt) in the system of real numbers, separating all rational numbers into two groups having certain characteristic properties. The subject has received later contributions at the hands of Weierstrass, Kronecker, and Méray.

The search for roots of quintic and higher degree equations was an important development, the Abel–Ruffini theorem (Ruffini 1799, Abel 1824) showed that they could not be solved by radicals (formulas involving only arithmetical operations and roots). Hence it was necessary to consider the wider set of algebraic numbers (all solutions to polynomial equations). Galois (1832) linked polynomial equations to group theory giving rise to the field of Galois theory.

Continued fractions, closely related to irrational numbers (and due to Cataldi, 1613), received attention at the hands of Euler, and at the opening of the 19th century were brought into prominence through the writings of Joseph Louis Lagrange. Other noteworthy contributions have been made by Druckenmüller (1837), Kunze (1857), Lemke (1870), and Günther (1872). Ramus first connected the subject with determinants, resulting, with the subsequent contributions of Heine, Möbius, and Günther, in the theory of Kettenbruchdeterminanten.

HISTORY OF IRRATIONAL NUMBER

In mathematics, the irrational numbers (from in- prefix assimilated to ir- (negative prefix, privative) + rational) are all the real numbers that are not rational numbers. That is, irrational numbers cannot be expressed as the ratio of two integers. When the ratio of lengths of two line segments is an irrational number, the line segments are also described as being incommensurable, meaning that they share no "measure" in common, that is, there is no length ("the measure"), no matter how short, that could be used to express the lengths of both of the two given segments as integer multiples of itself.

Among irrational numbers are the ratio π of a circle's circumference to its diameter, Euler's number e, the golden ratio ö, and the square root of two. In fact, all square roots of natural numbers, other than of perfect squares, are irrational. Like all real numbers, irrational numbers can be expressed in positional notation, notably as a decimal number. In the case of irrational numbers, the decimal expansion does not terminate, nor end with a repeating sequence. For example, the decimal representation of π starts with 3.14159, but no finite number of digits can represent π exactly, nor does it repeat. Conversely, a decimal expansion that terminates or repeats must be a rational number. These are provable properties of rational numbers and positional number systems, and are not used as definitions in mathematics.

Irrational numbers can also be expressed as non-terminating continued fractions and many other ways.

As a consequence of Cantor's proof that the real numbers are uncountable and the rationals countable, it follows that almost all real numbers are irrational.

Ancient Greece

The first proof of the existence of irrational numbers is usually attributed to a Pythagorean (possibly Hippasus of Metapontum), who probably discovered them while identifying sides of the pentagram. The then-current Pythagorean method would have claimed that there must be some sufficiently small, indivisible unit that could fit evenly into one of these lengths as well as the other. Hippasus, in the 5th century BC, however, was able to deduce that there was in fact no common unit of measure, and that the assertion of such an existence was in fact a contradiction. He did this by demonstrating that if the hypotenuse of an isosceles right triangle was indeed commensurable with a leg, then one of those lengths measured in that unit of measure must be both odd and even, which is impossible. His reasoning is as follows:

- Start with an isosceles right triangle with side lengths of integers a, b, and c. The ratio of the hypotenuse to a leg is represented by c:b.
- Assume a, b, and c are in the smallest possible terms (i.e. they have no common factors).
- By the Pythagorean theorem: $c^2 = a^2+b^2 = b^2+b^2 = 2b^2$. (Since the triangle is isosceles, a = b).
- Since $c^2 = 2b^2$, c^2 is divisible by 2, and therefore even.
- Since c^2 is even, c must be even.
- Since c is even, dividing c by 2 yields an integer. Let y be this integer ($c = 2y$).
- Squaring both sides of $c = 2y$ yields $c^2 = (2y)^2$, or $c^2 = 4y^2$.
- Substituting $4y^2$ for c^2 in the first equation ($c^2 = 2b^2$) gives us $4y^2= 2b^2$.
- Dividing by 2 yields $2y^2 = b^2$.

- Since y is an integer, and $2y^2 = b^2$, b^2 is divisible by 2, and therefore even.
- Since b^2 is even, b must be even.
- We have just shown that both b and c must be even. Hence they have a common factor of 2. However this contradicts the assumption that they have no common factors. This contradiction proves that c and b cannot both be integers, and thus the existence of a number that cannot be expressed as a ratio of two integers.

Greek mathematicians termed this ratio of incommensurable magnitudes alogos, or inexpressible. Hippasus, however, was not lauded for his efforts: according to one legend, he made his discovery while out at sea, and was subsequently thrown overboard by his fellow Pythagoreans "... for having produced an element in the universe which denied the... doctrine that all phenomena in the universe can be reduced to whole numbers and their ratios." Another legend states that Hippasus was merely exiled for this revelation. Whatever the consequence to Hippasus himself, his discovery posed a very serious problem to Pythagorean mathematics, since it shattered the assumption that number and geometry were inseparable–a foundation of their theory.

The discovery of incommensurable ratios was indicative of another problem facing the Greeks: the relation of the discrete to the continuous. This was brought into light by Zeno of Elea, who questioned the conception that quantities are discrete and composed of a finite number of units of a given size. Past Greek conceptions dictated that they necessarily must be, for "whole numbers represent discrete objects, and a commensurable ratio represents a relation between two collections of discrete objects", but Zeno found that in fact "[quantities] in general are not discrete collections of units; this is why ratios of incommensurable [quantities] appear... .[Q]uantities are, in other words, continuous". What this means is that, contrary to the popular conception of the time, there cannot be an indivisible, smallest unit of measure for any quantity. That in fact, these divisions of quantity must necessarily be infinite.

For example, consider a line segment: this segment can be split in half, that half split in half, the half of the half in half, and so on. This process can continue infinitely, for there is always another half to be split. The more times the segment is halved, the closer the unit of measure comes to zero, but it never reaches exactly zero. This is just what Zeno sought to prove. He sought to prove this by formulating four paradoxes, which demonstrated the contradictions inherent in the mathematical thought of the time. While Zeno's paradoxes accurately demonstrated the deficiencies of current mathematical conceptions, they were not regarded as proof of the alternative. In the minds of the Greeks, disproving the validity of one view did not necessarily prove the validity of another, and therefore further investigation had to occur.

The next step was taken by Eudoxus of Cnidus, who formalized a new theory of proportion that took into account commensurable as well as incommensurable quantities. Central to his idea was the distinction between magnitude and number. A magnitude "...was not a number but stood for entities such as line segments, angles, areas, volumes, and time which could vary, as we would say, continuously. Magnitudes were opposed to numbers, which jumped from one value to another, as from 4 to 5". Numbers are composed of some smallest, indivisible unit, whereas magnitudes are infinitely reducible.

Because no quantitative values were assigned to magnitudes, Eudoxus was then able to account for both commensurable and incommensurable ratios by defining a ratio in terms of its magnitude, and proportion as an equality between two ratios. By taking quantitative values (numbers) out of the equation, he avoided the trap of having to express an irrational number as a number. "Eudoxus' theory enabled the Greek mathematicians to make tremendous progress in geometry by supplying the necessary logical foundation for incommensurable ratios". This incommensurability is dealt with in Euclid's Elements, Book X, Proposition 9. It was not until Eudoxus developed a theory of proportion that took into account irrational as well as rational ratios that a strong mathematical foundation of irrational numbers was created.

As a result of the distinction between number and magnitude, geometry became the only method that could take into account incommensurable ratios. Because previous numerical foundations were still incompatible with the concept of incommensurability, Greek focus shifted away from those numerical conceptions such as algebra and focused almost exclusively on geometry. In fact, in many cases algebraic conceptions were reformulated into geometric terms. This may account for why we still conceive of x^2 and x^3 as x squared and x cubed instead of x to the second power and x to the third power.

Also crucial to Zeno's work with incommensurable magnitudes was the fundamental focus on deductive reasoning that resulted from the foundational shattering of earlier Greek mathematics. The realization that some basic conception within the existing theory was at odds with reality necessitated a complete and thorough investigation of the axioms and assumptions that underlie that theory. Out of this necessity, Eudoxus developed his method of exhaustion, a kind of reductio ad absurdum that "...established the deductive organization on the basis of explicit axioms..." as well as "...reinforced the earlier decision to rely on deductive reasoning for proof". This method of exhaustion is the first step in the creation of calculus.

Theodorus of Cyrene proved the irrationality of the surds of whole numbers up to 17, but stopped there probably because the algebra he used could not be applied to the square root of 17.

INDIA

Geometrical and mathematical problems involving irrational numbers such as square roots were addressed very early during the Vedic period in India. There are references to such calculations in the Samhitas, Brahmanas, and the Shulba Sutras (800 BC or earlier). (See Bag, Indian Journal of History of Science, 25(1-4), 1990).

It is suggested that the concept of irrationality was implicitly accepted by Indian mathematicians since the 7th century BC, when Manava (c. 750 – 690 BC) believed that the square roots of numbers such as 2 and 61 could not be exactly determined. Historian Carl Benjamin Boyer, however, writes that "such claims are not well substantiated and unlikely to be true".

It is also suggested that Aryabhata (5th century AD), in calculating a value of pi to 5 significant figures, used the word âsanna (approaching), to mean that not only is this an approximation but that the value is incommensurable (or irrational).

Later, in their treatises, Indian mathematicians wrote on the arithmetic of surds including addition, subtraction, multiplication, rationalization, as well as separation and extraction of square roots.

Mathematicians like Brahmagupta (in 628 AD) and Bhâṣkara I (in 629 AD) made contributions in this area as did other mathematicians who followed. In the 12th century Bhâskara II evaluated some of these formulas and critiqued them, identifying their limitations.

During the 14th to 16th centuries, Madhava of Sangamagrama and the Kerala school of astronomy and mathematics discovered the infinite series for several irrational numbers such as π and certain irrational values of trigonometric functions. Jyecmhadeva provided proofs for these infinite series in the Yuktibhâcâ.

Middle Ages

In the Middle ages, the development of algebra by Muslim mathematicians allowed irrational numbers to be treated as algebraic objects. Middle Eastern mathematicians also merged the concepts of "number" and "magnitude" into a more general idea of real numbers, criticized Euclid's idea of ratios, developed the theory of composite ratios, and extended the concept of number to ratios of continuous magnitude. In his commentary on Book 10 of the Elements, the Persian mathematician Al-Mahani (d. 874/884) examined and classified quadratic irrationals and cubic irrationals. He provided definitions for rational and irrational magnitudes, which he treated as irrational numbers. He dealt with them freely but explains them in geometric terms as follows:

"It will be a rational (magnitude) when we, for instance, say 10, 12, 3%, 6%, etc., because its value is pronounced and expressed quantitatively. What is not rational is irrational and it is impossible to pronounce and represent its value quantitatively. For example: the roots of numbers such as 10, 15, 20 which are not squares, the sides of numbers which are not cubes etc."

In contrast to Euclid's concept of magnitudes as lines, Al-Mahani considered integers and fractions as rational magnitudes, and square roots and cube roots as irrational magnitudes. He also introduced an arithmetical approach to the concept of irrationality, as he attributes the following to irrational magnitudes:

"their sums or differences, or results of their addition to a rational magnitude, or results of subtracting a magnitude of this kind from an irrational one, or of a rational magnitude from it."

The Egyptian mathematician Abû Kâmil Shujâ ibn Aslam (c. 850 – 930) was the first to accept irrational numbers as solutions to quadratic equations or as coefficients in an equation, often in the form of square roots, cube roots and fourth roots. In the 10th century, the Iraqi mathematician Al-Hashimi provided general proofs (rather than geometric demonstrations) for irrational numbers, as he considered multiplication, division, and other arithmetical functions. Iranian mathematician, Abû Ja'far al-Khâzin (900–971) provides a definition of rational and irrational magnitudes, stating that if a definite quantity is:

"contained in a certain given magnitude once or many times, then this (given) magnitude corresponds to a rational number. . . . Each time when this (latter) magnitude comprises a half, or a third, or a quarter of the given magnitude (of the unit), or, compared with (the unit), comprises three, five, or three fifths, it is a rational magnitude. And, in general, each magnitude that corresponds to this magnitude (i.e. to the unit), as one number to another, is rational. If, however, a magnitude cannot be represented as a multiple, a part (1/n), or parts (m/n) of a given magnitude, it is irrational, i.e. it cannot be expressed other than by means of roots."

Many of these concepts were eventually accepted by European mathematicians sometime after the Latin translations of the 12th century. Al-Hassâr, a Moroccan mathematician from Fez specializing in Islamic inheritance jurisprudence during the 12th century, first mentions the use of a fractional bar, where numerators and denominators are separated by a horizontal bar. In his discussion he writes, "..., for example, if you are told to write three-fifths and a third of a fifth, write thus, $\frac{3\ \ 1}{5\ \ 3}$. This same fractional notation appears soon after in the work of Leonardo Fibonacci in the 13th century.

Modern period

The 17th century saw imaginary numbers become a powerful tool in the hands of Abraham de Moivre, and especially of Leonhard Euler. The completion of the theory of complex numbers in the 19th century entailed the differentiation of irrationals into algebraic and transcendental numbers, the proof of the existence of transcendental numbers, and the resurgence of the scientific study of the theory of irrationals, largely ignored since Euclid. The year 1872 saw the publication of the theories of Karl Weierstrass (by his pupil Ernst Kossak), Eduard Heine (Crelle's Journal, 74), Georg Cantor (Annalen, 5), and Richard Dedekind. Méray had taken in 1869 the same point of departure as Heine, but the theory is generally referred to the year 1872. Weierstrass's method has been completely set forth by Salvatore Pincherle in 1880, and Dedekind's has received additional prominence through the author's later work (1888) and the endorsement by Paul Tannery (1894). Weierstrass, Cantor, and Heine base their theories on infinite series, while Dedekind founds his on the idea of a cut (Schnitt) in the system of all rational numbers, separating them into two groups having certain characteristic properties. The subject has received later contributions at the hands of Weierstrass, Leopold Kronecker (Crelle, 101), and Charles Méray.

Continued fractions, closely related to irrational numbers (and due to Cataldi, 1613), received attention at the hands of Euler, and at the opening of the 19th century were brought into prominence through the writings of Joseph-Louis Lagrange. Dirichlet also added to the general theory, as have numerous contributors to the applications of the subject.

Johann Heinrich Lambert proved (1761) that π cannot be rational, and that e^n is irrational if n is rational (unless $n = 0$). While Lambert's proof is often called incomplete, modern assessments support it as satisfactory, and in fact for its time it is unusually rigorous. Adrien-Marie Legendre (1794), after introducing the Bessel–Clifford function, provided a proof to show that π^2 is irrational, whence it follows immediately that π is irrational also. The existence of transcendental numbers was first established by Liouville (1844, 1851). Later, Georg Cantor (1873) proved their existence by a different method, which showed that every interval in the reals contains transcendental numbers. Charles Hermite (1873) first proved e transcendental, and Ferdinand von Lindemann (1882), starting from

Hermite's conclusions, showed the same for π. Lindemann's proof was much simplified by Weierstrass (1885), still further by David Hilbert (1893), and was finally made elementary by Adolf Hurwitz and Paul Gordan.

Types

- number theoretic distinction : transcendental/algebraic
- normal/ abnormal (non-normal)

Transcendental/algebraic

Almost all irrational numbers are transcendental and all real transcendental numbers are irrational (there are also complex transcendental numbers): the article on transcendental numbers lists several examples. So e^r and π^r are irrational for all nonzero rational r, and, e.g., e^π is irrational, too.

Irrational numbers can also be found within the countable set of real algebraic numbers (essentially defined as the real roots of polynomials with integer coefficients), i.e., as real solutions of polynomial equations

$$p(x) = a_n x^n + a_{n-1} x^{n-1} + \cdots + a_1 x + a_0 = 0 ,$$

where the coefficients a_i are integers and a n ‴ 0. Any rational root of this polynomial equation must be of the form r /s, where r is a divisor of a_0 and s is a divisor of a_n. If a real root x_0 of a polynomial p is not among these finitely many possibilities, it must be an irrational algebraic number. An exemplary proof for the existence of such algebraic irrationals is by showing that $x_0 = (2^{1/2} + 1)^{1/3}$ is an irrational root of a polynomial with integer coefficients: it satisfies (x^{3} ″ 1)^{2} = 2 and hence x^{6} ″ 2x^{3} ″ 1 = 0, and this latter polynomial has no rational roots (the only candidates to check are ±1, and x_0, being greater than 1, is neither of these), so x_0 is an irrational algebraic number.

Because the algebraic numbers form a subfield of the real numbers, many irrational real numbers can be constructed by combining transcendental and algebraic numbers. For example, 3π + 2, π + ″2 and e″3 are irrational (and even transcendental).

In constructive mathematics

In constructive mathematics, excluded middle is not valid, so it is not true that every real number is rational or irrational. Thus, the notion of an irrational number bifurcates into multiple distinct notions. One could take the traditional definition of an irrational number as a real number that is not rational. However, there is a second definition of an irrational number used in constructive mathematics, that a real number r is an irrational number if it is apart from every rational number, or equivalently, if the distance | r " q | between r and every rational number q is positive. This definition is stronger than the traditional definition of an irrational number. This second definition is used in Errett Bishop's proof that the square root of 2 is irrational.

Set of all irrationals

Since the reals form an uncountable set, of which the rationals are a countable subset, the complementary set of irrationals is uncountable.

Under the usual (Euclidean) distance function d (x, y) = | x " y |, the real numbers are a metric space and hence also a topological space. Restricting the Euclidean distance function gives the irrationals the structure of a metric space. Since the subspace of irrationals is not closed, the induced metric is not complete. Being a G-delta set—i.e., a countable intersection of open subsets—in a complete metric space, the space of irrationals is completely metrizable: that is, there is a metric on the irrationals inducing the same topology as the restriction of the Euclidean metric, but with respect to which the irrationals are complete. One can see this without knowing the aforementioned fact about G-delta sets: the continued fraction expansion of an irrational number defines a homeomorphism from the space of irrationals to the space of all sequences of positive integers, which is easily seen to be completely metrizable.

Furthermore, the set of all irrationals is a disconnected metrizable space. In fact, the irrationals equipped with the subspace topology have a basis of clopen sets so the space is zero-dimensional.

Transcendental numbers and reals

The existence of transcendental numbers was first established by Liouville (1844, 1851). Hermite proved in 1873 that e is transcendental and Lindemann proved in 1882 that π is transcendental. Finally, Cantor showed that the set of all real numbers is uncountably infinite but the set of all algebraic numbers is countably infinite, so there is an uncountably infinite number of transcendental numbers.

Infinity and infinitesimals

The earliest known conception of mathematical infinity appears in the Yajur Veda, an ancient Indian script, which at one point states, "If you remove a part from infinity or add a part to infinity, still what remains is infinity." Infinity was a popular topic of philosophical study among the Jain mathematicians c. 400 BC. They distinguished between five types of infinity: infinite in one and two directions, infinite in area, infinite everywhere, and infinite perpetually. The symbol " is often used to represent an infinite quantity. Aristotle defined the traditional Western notion of mathematical infinity. He distinguished between actual infinity and potential infinity—the general consensus being that only the latter had true value. Galileo Galilei's Two New Sciences discussed the idea of one-to-one correspondences between infinite sets. But the next major advance in the theory was made by Georg Cantor; in 1895 he published a book about his new set theory, introducing, among other things, transfinite numbers and formulating the continuum hypothesis.

In the 1960s, Abraham Robinson showed how infinitely large and infinitesimal numbers can be rigorously defined and used to develop the field of nonstandard analysis. The system of hyperreal numbers represents a rigorous method of treating the ideas about infinite and infinitesimal numbers that had been used casually by mathematicians, scientists, and engineers ever since the invention of infinitesimal calculus by Newton and Leibniz.

A modern geometrical version of infinity is given by projective geometry, which introduces "ideal points at infinity", one for each spatial direction. Each family of parallel lines in a given direction is postulated to converge to the corresponding ideal point. This is closely related to the idea of vanishing points in perspective drawing.

COMPLEX NUMBERS

The earliest fleeting reference to square roots of negative numbers occurred in the work of the mathematician and inventor Heron of Alexandria in the 1st century AD, when he considered the volume of an impossible frustum of a pyramid. They became more prominent when in the 16th century closed formulas for the roots of third and fourth degree polynomials were discovered by Italian mathematicians such as Niccolò Fontana Tartaglia and Gerolamo Cardano. It was soon realized that these formulas, even if one was only interested in real solutions, sometimes required the manipulation of square roots of negative numbers.

This was doubly unsettling since they did not even consider negative numbers to be on firm ground at the time. When René Descartes coined the term "imaginary" for these quantities in 1637, he intended it as derogatory. (See imaginary number for a discussion of the "reality" of complex numbers.) A further source of confusion was that the equation

$$\left(\sqrt{-1}\right)^2 = \sqrt{-1}\sqrt{-1} = -1$$

seemed capriciously inconsistent with the algebraic identity

$$\sqrt{a}\sqrt{b} = \sqrt{ab},$$

which is valid for positive real numbers a and b, and was also used in complex number calculations with one of a, b positive and the other negative. The incorrect use of this identity, and the related identity

$$\frac{1}{\sqrt{a}} = \sqrt{\frac{1}{a}}$$

in the case when both a and b are negative even bedeviled Euler. This difficulty eventually led him to the convention of using the special symbol i in place of “1 to guard against this mistake.

The 18th century saw the work of Abraham de Moivre and Leonhard Euler. De Moivre's formula (1730) states:

$$(\cos\theta + i\sin\theta)^n = \cos n\theta + i\sin n\theta$$

while Euler's formula of complex analysis (1748) gave us:

$$\cos\theta + i\sin\theta = e^{i\theta}.$$

The existence of complex numbers was not completely accepted until Caspar Wessel described the geometrical interpretation in 1799. Carl Friedrich Gauss rediscovered and popularized it several years later, and as a result the theory of complex numbers received a notable expansion. The idea of the graphic representation of complex numbers had appeared, however, as early as 1685, in Wallis's De algebra tractatus.

Also in 1799, Gauss provided the first generally accepted proof of the fundamental theorem of algebra, showing that every polynomial over the complex numbers has a full set of solutions in that realm. The general acceptance of the theory of complex numbers is due to the labors of Augustin Louis Cauchy and Niels Henrik Abel, and especially the latter, who was the first to boldly use complex numbers with a success that is well known.[peacock prose]

Gauss studied complex numbers of the form a + bi, where a and b are integral, or rational (and i is one of the two roots of $x^2 + 1 = 0$). His student, Gotthold Eisenstein, studied the type a + bù, where ù is a complex root of x^3 “ 1 = 0. Other such classes (called cyclotomic fields) of complex numbers derive from the roots of unity x^k “ 1 = 0 for higher values of k. This generalization is largely due to Ernst Kummer, who also invented ideal numbers, which were expressed as geometrical entities by Felix Klein in 1893.

In 1850 Victor Alexandre Puiseux took the key step of distinguishing between poles and branch points, and introduced the concept of essential singular points. This eventually led to the concept of the extended complex plane.

PRIME NUMBERS

Prime numbers have been studied throughout recorded history. Euclid devoted one book of the Elements to the theory of primes; in it he proved the infinitude of the primes and the fundamental theorem of arithmetic, and presented the Euclidean algorithm for finding the greatest common divisor of two numbers.

In 240 BC, Eratosthenes used the Sieve of Eratosthenes to quickly isolate prime numbers. But most further development of the theory of primes in Europe dates to the Renaissance and later eras.

In 1796, Adrien-Marie Legendre conjectured the prime number theorem, describing the asymptotic distribution of primes. Other results concerning the distribution of the primes include Euler's proof that the sum of the reciprocals of the primes diverges, and the Goldbach conjecture, which claims that any sufficiently large even number is the sum of two primes. Yet another conjecture related to the distribution of prime numbers is the Riemann hypothesis, formulated by Bernhard Riemann in 1859. The prime number theorem was finally proved by Jacques Hadamard and Charles de la Vallée-Poussin in 1896. Goldbach and Riemann's conjectures remain unproven and unrefuted.

MAIN CLASSIFICATION

Numbers can be classified into sets, called number systems, such as the natural numbers and the real numbers. The major categories of numbers are as follows:

There is generally no problem in identifying each number system with a proper subset of the next one (by abuse of notation), because each of these number systems is canonically isomorphic to a proper subset of the next one. The resulting hierarchy allows, for example, to talk, formally correctly, about real numbers that are rational numbers, and is expressed symbolically by writing

$$\mathbb{N} \subset \mathbb{Z} \subset \mathbb{Q} \subset \mathbb{R} \subset \mathbb{C}.$$

Natural numbers

The most familiar numbers are the natural numbers (sometimes called whole numbers or counting numbers): 1, 2, 3, and so on. Traditionally, the sequence of natural numbers started with 1 (0 was not even considered a number for the Ancient Greeks.) However, in the 19th century, set theorists and other mathematicians started including 0 (cardinality of the empty set, i.e. 0 elements, where 0 is thus the smallest cardinal number) in the set of natural numbers. Today, different mathematicians use the term to describe both sets, including 0 or not. The mathematical symbol for the set of all natural numbers is N, also written N, and sometimes N_0 or N_1 when it is necessary to indicate whether the set should start with 0 or 1, respectively.

In the base 10 numeral system, in almost universal use today for mathematical operations, the symbols for natural numbers are written using ten digits: 0, 1, 2, 3, 4, 5, 6, 7, 8, and 9. The radix or base is the number of unique numerical digits, including zero, that a numeral system uses to represent numbers (for the decimal system, the radix is 10). In this base 10 system, the rightmost digit of a natural number has a place value of 1, and every other digit has a place value ten times that of the place value of the digit to its right.

In set theory, which is capable of acting as an axiomatic foundation for modern mathematics, natural numbers can be represented by classes of equivalent sets. For instance, the number 3 can be represented as the class of all sets that have exactly three elements. Alternatively, in Peano Arithmetic, the number 3 is represented as sss0, where s is the "successor" function (i.e., 3 is the third successor of 0). Many different representations are possible; all that is needed to formally represent 3 is to inscribe a certain symbol or pattern of symbols three times.

Integers

The negative of a positive integer is defined as a number that produces 0 when it is added to the corresponding positive integer. Negative numbers are usually written with a negative sign (a minus sign). As an example, the negative of 7 is written "7, and 7 + ("7) = 0. When the set of negative numbers is combined with the set of natural numbers (including 0), the result is defined as the set of integers, Z also written Z . Here the letter Z comes from German Zahl 'number'. The set of integers forms a ring with the operations addition and multiplication.

The natural numbers form a subset of the integers. As there is no common standard for the inclusion or not of zero in the natural numbers, the natural numbers without zero are commonly referred to as positive integers, and the natural numbers with zero are referred to as non-negative integers.

Rational numbers

A rational number is a number that can be expressed as a fraction with an integer numerator and a positive integer denominator. Negative denominators are allowed, but are commonly avoided, as every rational number is equal to a fraction with positive denominator. Fractions are written as two integers, the numerator and the denominator, with a dividing bar between them. The fraction m/n represents m parts of a whole divided into n equal parts. Two different fractions may correspond to the same rational number; for example 1/2 and 2/4 are equal, that is:

$$\frac{1}{2}=\frac{2}{4}.$$

In general,

$$\frac{a}{b}=\frac{c}{d} \text{ if and only if } a \times d = c \times b.$$

If the absolute value of m is greater than n (supposed to be positive), then the absolute value of the fraction is greater than 1. Fractions can be greater than, less than, or equal to 1 and can also be positive, negative, or 0. The set of all rational numbers includes the integers since every integer can be written as a fraction with denominator 1. For example "7 can be written "7/1. The symbol for the rational numbers is Q (for quotient), also written Q.

Real numbers

The symbol for the real numbers is R, also written as R. They include all the measuring numbers. Every real number corresponds to a point on the number line. The following paragraph will focus primarily on positive real numbers. The treatment of negative real numbers is according to the general rules of arithmetic and their denotation is simply prefixing the corresponding positive numeral by a minus sign, e.g. "123.456.

Most real numbers can only be approximated by decimal numerals, in which a decimal point is placed to the right of the digit with place value 1. Each digit to the right of the decimal point has a place value one-tenth of the place value of the digit to its left. For example, 123.456 represents 123456/1000, or, in words, one hundred, two tens, three ones, four tenths, five hundredths, and six thousandths. A real number can be expressed by a finite number of decimal digits only if it is rational and its fractional part has a denominator whose prime factors are 2 or 5 or both, because these are the prime factors of 10, the base of the decimal system.

Thus, for example, one half is 0.5, one fifth is 0.2, one-tenth is 0.1, and one fiftieth is 0.02. Representing other real numbers as decimals would require an infinite sequence of digits to the right of the decimal point. If this infinite sequence of digits follows a pattern, it can be written with an ellipsis or another notation that indicates the repeating pattern. Such a decimal is called a repeating decimal. Thus 1/3 can be written as 0.333..., with an ellipsis to indicate that the pattern continues. Forever repeating 3s are also written as 0.3.

It turns out that these repeating decimals (including the repetition of zeroes) denote exactly the rational numbers, i.e., all rational numbers are also real numbers, but it is not the case that every real number is rational. A real number that is not rational is called irrational. A famous irrational real number is the number π, the ratio of the circumference of any circle to its diameter. When pi is written as

$$\pi = 3.14159265358979\ldots,$$

as it sometimes is, the ellipsis does not mean that the decimals repeat (they do not), but rather that there is no end to them. It has been proved that π is irrational. Another well-known number, proven to be an irrational real number, is

$$\sqrt{2} = 1.41421356237\ldots,$$

the square root of 2, that is, the unique positive real number whose square is 2. Both these numbers have been approximated (by computer) to trillions (1 trillion = 10^{12} = 1,000,000,000,000) of digits.

Not only these prominent examples but almost all real numbers are irrational and therefore have no repeating patterns and hence no corresponding decimal numeral. They can only be approximated by decimal numerals, denoting rounded or truncated real numbers. Any rounded or truncated number is necessarily a rational number, of which there are only countably many. All measurements are, by their nature, approximations, and always have a margin of error. Thus 123.456 is considered an approximation of any real number greater or equal to 1234555/10000 and strictly less than 1234565/10000 (rounding to 3 decimals), or of any real number greater or equal to 123456/1000 and strictly less than 123457/1000 (truncation after the 3. decimal).

Digits that suggest a greater accuracy than the measurement itself does, should be removed. The remaining digits are then called significant digits. For example, measurements with a ruler can seldom be made without a margin of error of at least 0.001 m. If the sides of a rectangle are measured as 1.23 m and 4.56 m, then multiplication gives an area for the rectangle between 5.614591 m^2 and 5.603011 m^2. Since not even the second digit after the decimal place is preserved, the following digits are not significant. Therefore, the result is usually rounded to 5.61.

Just as the same fraction can be written in more than one way, the same real number may have more than one decimal representation. For example, 0.999..., 1.0, 1.00, 1.000, ..., all represent the natural number 1. A given real number has only the following decimal representations: an approximation to some finite number of decimal places, an approximation in which a pattern is established that continues for an unlimited number of decimal places or an exact value with only finitely many decimal places. In this last case, the last non-zero digit may be replaced by the digit one smaller followed by an unlimited number of 9's, or the last non-zero digit may be followed by an unlimited number of zeros.

Thus the exact real number 3.74 can also be written 3.7399999999... and 3.74000000000.... Similarly, a decimal numeral with an unlimited number of 0's can be rewritten by dropping the 0's to the right of the decimal place, and a decimal numeral with an unlimited number of 9's can be rewritten by increasing the rightmost -9 digit by one, changing all the 9's to the right of that digit to 0's. Finally, an unlimited sequence of 0's to the right of the decimal place can be dropped. For example, 6.849999999999... = 6.85 and 6.850000000000... = 6.85. Finally, if all of the digits in a numeral are 0, the number is 0, and if all of the digits in a numeral are an unending string of 9's, you can drop the nines to the right of the decimal place, and add one to the string of 9s to the left of the decimal place. For example, 99.999... = 100.

The real numbers also have an important but highly technical property called the least upper bound property.

It can be shown that any ordered field, which is also complete, is isomorphic to the real numbers. The real numbers are not, however, an algebraically closed field, because they do not include a solution (often called a square root of minus one) to the algebraic equation $x^2 + 1 = 0$.

Complex numbers

Moving to a greater level of abstraction, the real numbers can be extended to the complex numbers. This set of numbers arose historically from trying to find closed formulas for the roots of cubic and quadratic polynomials. This led to expressions involving the square roots of negative numbers, and eventually to the definition of a new number: a square root of "1, denoted by i, a symbol assigned by Leonhard Euler, and called the imaginary unit. The complex numbers consist of all numbers of the form

$$a + bi$$

where a and b are real numbers. Because of this, complex numbers correspond to points on the complex plane, a vector space of two real dimensions. In the expression a + bi, the real number a is called the real part and b is called the imaginary part. If the real part of a complex number is 0, then the number is called an imaginary number or is referred to as purely imaginary; if the imaginary part is 0, then the number is a real number. Thus the real numbers are a subset of the complex numbers. If the real and imaginary parts of a complex number are both integers, then the number is called a Gaussian integer. The symbol for the complex numbers is C or C.

The fundamental theorem of algebra asserts that the complex numbers form an algebraically closed field, meaning that every polynomial with complex coefficients has a root in the complex numbers. Like the reals, the complex numbers form a field, which is complete, but unlike the real numbers, it is not ordered. That is, there is no consistent meaning assignable to saying that i is greater than 1, nor is there any meaning in saying that i is less than 1. In technical terms, the complex numbers lack a total order that is compatible with field operations.

SUBCLASSES OF THE INTEGERS

Even and odd numbers

An even number is an integer that is "evenly divisible" by two, that is divisible by two without remainder; an odd number is an integer that is not even. (The old-fashioned term "evenly divisible" is now almost always shortened to "divisible".) Any odd number n may be constructed by the formula $n = 2k + 1$, for a suitable integer k. Starting with $k = 0$, the first non-negative odd numbers are {1, 3, 5, 7, ...}. Any even number m has the form $m = 2k$ where k is again an integer. Similarly, the first non-negative even numbers are {0, 2, 4, 6, ...}.

Prime numbers

A prime number, often shortened to just prime, is an integer greater than 1 that is not the product of two smaller positive integers. The first few prime numbers are 2, 3, 5, 7, and 11. There is no such simple formula as for odd and even numbers to generate the prime numbers. The primes have been widely studied for more than 2000 years and have led to many questions, only some of which have been answered. The study of these questions belongs to number theory. Goldbach's conjecture is an example of a still unanswered question: "Is every even number the sum of two primes?"

One answered question, as to whether every integer greater than one is a product of primes in only one way, except for a rearrangement of the primes, was confirmed; this proven claim is called the fundamental theorem of arithmetic. A proof appears in Euclid's Elements.

Other classes of integers

Many subsets of the natural numbers have been the subject of specific studies and have been named, often after the first mathematician that has studied them. Example of such sets of integers are Fibonacci numbers and perfect numbers. For more examples, see Integer sequence.

Subclasses of the complex numbers

Algebraic, irrational and transcendental numbers

Algebraic numbers are those that are a solution to a polynomial equation with integer coefficients. Real numbers that are not rational numbers are called irrational numbers. Complex numbers which are not algebraic are called transcendental numbers. The algebraic numbers that are solutions of a monic polynomial equation with integer coefficients are called algebraic integers.

Constructible numbers

Motivated by the classical problems of constructions with straightedge and compass, the constructible numbers are those complex numbers whose real and imaginary parts can be constructed using straightedge and compass, starting from a given segment of unit length, in a finite number of steps.

Computable numbers

A computable number, also known as recursive number, is a real number such that there exists an algorithm which, given a positive number n as input, produces the first n digits of the computable number's decimal representation. Equivalent definitions can be given using ì-recursive functions, Turing machines or ë-calculus. The computable numbers are stable for all usual arithmetic operations, including the computation of the roots of a polynomial, and thus form a real closed field that contains the real algebraic numbers.

The computable numbers may be viewed as the real numbers that may be exactly represented in a computer: a computable number is exactly represented by its first digits and a program for computing further digits. However, the computable numbers are rarely used in practice. One reason is that there is no algorithm for testing the equality of two computable numbers. More precisely, there cannot exist any algorithm which takes any computable number as an input, and decides in every case if this number is equal to zero or not.

The set of computable numbers has the same cardinality as the natural numbers. Therefore, almost all real numbers are non-computable. However, it is very difficult to produce explicitly a real number that is not computable.

Extensions of the concept

p-adic numbers

The p-adic numbers may have infinitely long expansions to the left of the decimal point, in the same way that real numbers may have infinitely long expansions to the right. The number system that results depends on what base is used for the digits: any base is possible, but a prime number base provides the best mathematical properties. The set of the p-adic numbers contains the rational numbers, but is not contained in the complex numbers.

The elements of an algebraic function field over a finite field and algebraic numbers have many similar properties. Therefore, they are often regarded as numbers by number theorists. The p-adic numbers play an important role in this analogy.

HYPERCOMPLEX NUMBERS

Some number systems that are not included in the complex numbers may be constructed from the real numbers in a way that generalize the construction of the complex numbers. They are sometimes called hypercomplex numbers. They include the quaternions H, introduced by Sir William Rowan Hamilton, in which multiplication is not commutative, the octonions, in which multiplication is not associative in addition to not being commutative, and the sedenions, in which multiplication is not alternative, neither associative nor commutative.

Transfinite numbers

For dealing with infinite sets, the natural numbers have been generalized to the ordinal numbers and to the cardinal numbers. The former gives the ordering of the set, while the latter gives its size. For finite sets, both ordinal and cardinal numbers are identified with the natural numbers. In the infinite case, many ordinal numbers correspond to the same cardinal number.

Nonstandard numbers

Hyperreal numbers are used in non-standard analysis. The hyperreals, or nonstandard reals (usually denoted as *R), denote an ordered field that is a proper extension of the ordered field of real numbers R and satisfies the transfer principle. This principle allows true first-order statements about R to be reinterpreted as true first-order statements about *R.

Superreal and surreal numbers extend the real numbers by adding infinitesimally small numbers and infinitely large numbers, but still form fields.

Fractions and Decimals

Fractions are a part of a whole. They are represented by numbers that have two parts to them and they are separated by a fraction bar. There is a number at the top, which is called the numerator, and the number at the bottom is called the denominator.

Example: The following figure shows a fraction in which 1 is the numerator and 4 is the denominator.

MULTIPLICATIVE INVERSES

The multiplicative inverse is defined as the reciprocal of a given number. It is used to simplify mathematical expressions. The word 'inverse' implies something opposite/contrary in effect, order, position, or direction. A number when multiplied to its multiplicative inverse results in 1.

In mathematics, a multiplicative inverse or reciprocal for a number x, denoted by 1/x or $x^{"1}$, is a number which when multiplied by x yields the multiplicative identity, 1. The multiplicative inverse of a fraction a/b is b/a. For the multiplicative inverse of a real number, divide 1 by the number. For example, the reciprocal of 5 is one fifth (1/5 or 0.2), and the reciprocal of 0.25 is 1 divided by 0.25, or 4. The reciprocal function, the function f(x) that maps x to 1/x, is one of the simplest examples of a function which is its own inverse (an involution).

Multiplying by a number is the same as dividing by its reciprocal and vice versa. For example, multiplication by 4/5 (or 0.8) will give the same result as division by 5/4 (or 1.25). Therefore, multiplication by a number followed by multiplication by its reciprocal yields the original number (since the product of the number and its reciprocal is 1).

The term reciprocal was in common use at least as far back as the third edition of Encyclopædia Britannica (1797) to describe two numbers whose product is 1; geometrical quantities in inverse proportion are described as reciprocall in a 1570 translation of Euclid's Elements.

In the phrase multiplicative inverse, the qualifier multiplicative is often omitted and then tacitly understood (in contrast to the additive inverse). Multiplicative inverses can be defined over many mathematical domains as well as numbers. In these cases it can happen that ab '" ba; then "inverse" typically implies that an element is both a left and right inverse.

The notation $f^{"1}$ is sometimes also used for the inverse function of the function f, which is not in general equal to the multiplicative inverse. For example, the multiplicative inverse 1/(sin x) = $(\sin x)^{"1}$ is the cosecant of x, and not the inverse sine of x denoted by $\sin^{"1}$ x or arcsin x. Only for linear maps are they strongly related. The terminology difference reciprocal versus inverse is not sufficient to make this distinction, since many authors prefer the opposite naming convention, probably for historical reasons (for example in French, the inverse function is preferably called the bijection réciproque).

Examples and counterexamples

In the real numbers, zero does not have a reciprocal because no real number multiplied by 0 produces 1 (the product of any number with zero is zero). With the exception of zero, reciprocals of every real number are real, reciprocals of every rational number are rational, and reciprocals of every complex number are complex. The property that every element other than zero has a multiplicative inverse is part of the definition of a field, of which these are all examples. On the other hand, no integer other than 1 and "1 has an integer reciprocal, and so the integers are not a field.

In modular arithmetic, the modular multiplicative inverse of a is also defined: it is the number x such that ax a" 1 (mod n). This multiplicative inverse exists if and only if a and n are coprime. For example, the inverse of 3 modulo 11 is 4 because 4 Å" 3 a" 1 (mod 11). The extended Euclidean algorithm may be used to compute it.

The sedenions are an algebra in which every nonzero element has a multiplicative inverse, but which nonetheless has divisors of zero, that is, nonzero elements x, y such that xy = 0.

A square matrix has an inverse if and only if its determinant has an inverse in the coefficient ring. The linear map that has the matrix A"1 with respect to some base is then the reciprocal function of the map having A as matrix in the same base. Thus, the two distinct notions of the inverse of a function are strongly related in this case, while they must be carefully distinguished in the general case (as noted above).

The trigonometric functions are related by the reciprocal identity: the cotangent is the reciprocal of the tangent; the secant is the reciprocal of the cosine; the cosecant is the reciprocal of the sine.

A ring in which every nonzero element has a multiplicative inverse is a division ring; likewise an algebra in which this holds is a division algebra.

Complex numbers

As mentioned above, the reciprocal of every nonzero complex number z = a + bi is complex. It can be found by multiplying both top and bottom of 1/z by its complex conjugate $\overline{z} = a - bi$ and using the property that $z\overline{z} = \| z \|^2$, the absolute value of z squared, which is the real number a2 + b2:

$$\frac{1}{z} = \frac{\overline{z}}{z\overline{z}} = \frac{\overline{z}}{\| z \|^2} = \frac{a-bi}{a^2+b^2} = \frac{a}{a^2+b^2} - \frac{b}{a^2+b^2}i.$$

The intuition is that

$$\frac{\bar{z}}{\| z \|}$$

gives us the complex conjugate with a magnitude reduced to a value of 1, so dividing again by |z| ensures that the magnitude is now equal to the reciprocal of the original magnitude as well, hence:

$$\frac{1}{z} = \frac{\bar{z}}{\| z \|^2}$$

In particular, if ||z||=1 (z has unit magnitude), then 1/z = z. Consequently, the imaginary units, ±i, have additive inverse equal to multiplicative inverse, and are the only complex numbers with this property. For example, additive and multiplicative inverses of i are "(i) = "i and 1/i = "i, respectively.

For a complex number in polar form z = r(cos ö + i sin ö), the reciprocal simply takes the reciprocal of the magnitude and the negative of the angle:

$$\frac{1}{z} = \frac{1}{r}\left(\cos(-\varphi) + i\sin(-\varphi)\right).$$

Algorithms

The reciprocal may be computed by hand with the use of long division.

Computing the reciprocal is important in many division algorithms, since the quotient a/b can be computed by first computing 1/b and then multiplying it by a. Noting that f(x) = 1/x "b has a zero at x = 1/b, Newton's method can find that zero, starting with a guess x_0 and iterating using the rule:

$$x_{n+1} = x_n - \frac{f(x_n)}{f'(x_n)} = x_n - \frac{1/x_n - b}{-1/x_n^2} = 2x_n - bx_n^2 = x_n(2 - bx_n).$$

This continues until the desired precision is reached. For example, suppose we wish to compute 1/17 H" 0.0588 with 3 digits of precision. Taking x_0 = 0.1, the following sequence is produced:

x_1 = 0.1(2 " 17 × 0.1) = 0.03

x_2 = 0.03(2 " 17 × 0.03) = 0.0447

x_3 = 0.0447(2 " 17 × 0.0447) H" 0.0554

x_4 = 0.0554(2 " 17 × 0.0554) H" 0.0586

x_5 = 0.0586(2 " 17 × 0.0586) H" 0.0588

A typical initial guess can be found by rounding b to a nearby power of 2, then using bit shifts to compute its reciprocal.

In constructive mathematics, for a real number x to have a reciprocal, it is not sufficient that x ‴ 0. There must instead be given a rational number r such that $0 < r < |x|$. In terms of the approximation algorithm described above, this is needed to prove that the change in y will eventually become arbitrarily small.

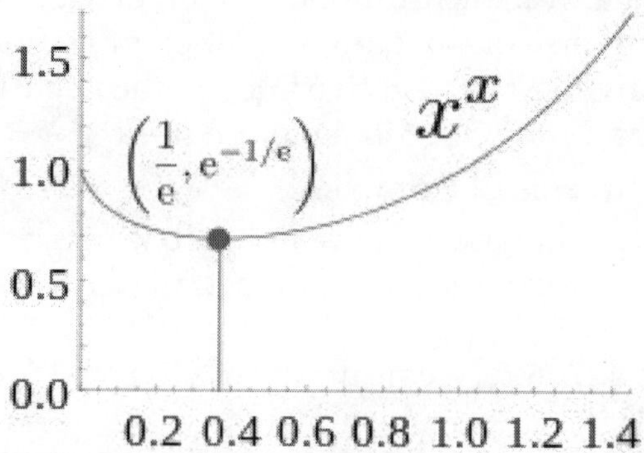

Graph of $f(x) = x^x$ showing the minimum at $(1/e, e^{''1/e})$.

This iteration can also be generalized to a wider sort of inverses; for example, matrix inverses.

Further remarks®

If the multiplication is associative, an element x with a multiplicative inverse cannot be a zero divisor (x is a zero divisor if some nonzero y, xy = 0). To see this, it is sufficient to multiply the equation xy = 0 by the inverse of x (on the left), and then simplify using associativity. In the absence of associativity, the sedenions provide a counterexample.

The converse does not hold: an element which is not a zero divisor is not guaranteed to have a multiplicative inverse. Within Z, all integers except "1, 0, 1 provide examples; they are not zero divisors nor do they have inverses in Z. If the ring or algebra is finite, however, then all elements a which are not zero divisors do have a (left and right) inverse. For, first observe that the map f(x) = ax must be injective: f(x) = f(y) implies x = y:

$$\begin{aligned} ax = ay \quad &\Rightarrow \quad ax - ay = 0 \\ &\Rightarrow \quad a(x-y) = 0 \\ &\Rightarrow \quad x - y = 0 \\ &\Rightarrow \quad x = y. \end{aligned}$$

Distinct elements map to distinct elements, so the image consists of the same finite number of elements, and the map is necessarily surjective. Specifically, *f* (namely multiplication by a) must map some element x to 1, ax = 1, so that x is an inverse for a.

Multiplicative Inverse of Integers

Finding the multiplicative inverse of positive integers is the same as natural numbers (explained above). Just like positive integers, the product of a negative number and its reciprocal must be equal to 1. Thus, the multiplicative inverse of any negative number is its reciprocal. For example, (-6) × (-1/6) = 1, therefore, the multiplicative inverse of -6 is -1/6. Note that the multiplicative inverse of a negative number is always negative. And, in the multiplicative inverse of a negative integer, the negative sign will attach to the numerator, and not with the denominator.

Multiplicative Inverse of a Fraction

The multiplicative inverse of a fraction a/b is b/a because a/b × b/a = 1 when (a,b ‴ 0). For example, the multiplicative inverse of 2/7 is 7/2. If we multiply 2/7 by 7/2, the product is 1 (2/7 × 7/2 = 1). The multiplicative inverse of 76/43 is 43/76. If we multiply 76/43 by 43/76, the product is 1 (76/43 × 43/76 = 1).

A unit fraction is a fraction with the numerator 1. If we multiply a unit fraction 1/x by x, the product is 1. Thus, the multiplicative inverse of a unit fraction 1/x is x.

Examples:

- The multiplicative inverse of the unit fraction 1/7 is 7. If we multiply 1/7 by 7, the product is 1 (1/7 × 7 = 1).
- The multiplicative inverse of the unit fraction 1/50 is 50. If we multiply 1/50 by 50, the product is 1 (1/50 × 50 = 1).

Multiplicative Inverse of a Mixed Fraction

To find the multiplicative inverse of a mixed fraction, convert the mixed fraction into an improper fraction, then determine its reciprocal. For example, let us find the multiplicative inverse of 312

- Step 1: Convert 3 1/2 to an improper fraction, that is 7/2.

Step 2: Find the reciprocal of 7/2, that is 2/7. Thus, the multiplicative inverse of 3 1/2 is 2/7.

It is interesting to note that the multiplicative inverse of a mixed number is always a proper fraction whose value is less than 1.

Multiplicative Inverse of 0

As per the definition of multiplicative inverse, it is the number that when multiplied to the original number results in 1 as the product. But with 0, we know that the product of 0 with any number is always 0. So, the multiplicative inverse of 0 does not exist.

We can also understand this using the properties of division which states that the division of any number by 0 is not defined. The multiplicative inverse of 0 can be written as 1/0, but its value is not defined. So, it does not exist.

Multiplicative Inverse Property

The multiplicative inverse property states that the product of a number with its reciprocal is always equal to 1. Look at the image given below where 1/n is the multiplicative inverse of the number n and 1 is the product.

For example, let us consider 5 apples. Now, divide the apples into five groups of 1 each. To make them into groups of 1 each, we need to divide them by 5. Dividing a number by itself is equivalent to multiplying it by its multiplicative inverse. Hence, $5 \div 5 = 5 \times 1/5 = 1$. Here, 1/5 is the multiplicative inverse of 5.

How to Find Multiplicative Inverse?

The multiplicative inverse of a number is also known as its reciprocal. It is very easy to find the multiplicative inverse of a number using the following steps:

- Step 1: Divide the given number by 1.
- Step 2: Write it in the form of a fraction. Say, the reciprocal of a is 1/a.
- Step 3: Simplify and get the answer.

Let us find the multiplicative inverse of 2/3. The first step is to divide it by 1, which will result in 1/(2/3) = 3/2. Therefore, the reciprocal of 2/3 is 3/2.

MULTIPLICATIVE INVERSE OF COMPLEX NUMBERS

Complex numbers of the form Z = a + ib, such as Z = 3+i″2, where 3 is the real number and i″2 is the imaginary number. The multiplicative inverse of a complex number Z is 1/Z. The reciprocal of this complex number is 1/ 3+i√2. It can be simplified by multiplying and dividing it by 3-i√2, such that: (3-i√2)/(3+i√2)(3-i√2) = (3-i√2)/(9-$i^2$2) = (3-i√2)/(9+2) = (3-i√2)/11. Therefore, 3/11 - i√2/11 is the multiplicative inverse of 3+i√2.

Follow the steps given below to find the multiplicative inverse of a complex number a + ib:

- Step 1: Write the reciprocal in the form of 1/(a+ib).
- Step 2: Multiply and divide this number by the conjugate of (a+ib).
- Step 3: Apply the following formulae: $(a + b)(a - b) = a^2 - b^2$, and $i^2 = -1$.
- Step 4: Simplify to the lowest form.

Modular Multiplicative Inverse

The modular multiplicative inverse of an integer p is another integer x such that the product px is congruent to 1 with respect to the modulus m. It can be represented as: px a″

1 (mod m). In other words, m divides px - 1 completely. Also, the modular multiplicative inverse of an integer p can exist with respect to the modulus m only if gcd(p, m) = 1.

In a nutshell, the multiplicative inverses are as follows:

Type	Multiplicative Inverse	Example
Natural Number x	1/x	Multiplicative Inverse of 4 is 1/4
Integer $x, x \neq 0$	1/x	Multiplicative Inverse of -4 is -1/4
Fraction $x/y; x,y \neq 0$	y/x	Multiplicative Inverse of 2/7 is 7/2
Unit Fraction $1/x, x \neq 0$	x	Multiplicative Inverse of 1/20 is 20

Tips and Tricks:

- The multiplicative inverse of a fraction can be obtained by flipping the numerator and denominator.
- The multiplicative inverse of 1 is 1.
- The multiplicative inverse of 0 is not defined.
- The multiplicative inverse of a number x is written as 1/x or x^{-1}.
- The multiplicative inverse of a mixed fraction can be obtained by converting the mixed fraction into an improper fraction and determining its reciprocal.

Important Notes

- The multiplicative inverse of a number is also called it's reciprocal.
- The product of a number and its multiplicative inverse is equal to 1.

Factors and Multiples

Factors and multiples are the two key concepts that are studied together. Factors are the numbers that divide the given number completely without leaving any remainder, whereas the multiples are the numbers that are multiplied by the other number to get specific numbers.

Factors of a given number are numbers that can perfectly divide that given number.

Examples:

- Factors of 6: 1, 2, 3, 6
- Factors of 8: 1, 2, 4, 8
- Factors of 14: 1, 2, 7, 14

- Factors of 36: 1, 2, 3, 4, 6, 9, 18, 36

A multiple of a number is a number obtained by multiplying the given number by another whole number.

Examples:

- Multiples of 3: 3, 6, 9, 12, 15, ...
- Multiples of 5: 5, 10, 15, 20, 25, ...
- Multiples of 10: 10, 20, 30, 40, 50,...
- Multiples of 12: 12, 24, 36, 48, 60, ...

Applications

The expansion of the reciprocal 1/q in any base can also act as a source of pseudo-random numbers, if q is a "suitable" safe prime, a prime of the form 2p + 1 where p is also a prime. A sequence of pseudo-random numbers of length q " 1 will be produced by the expansion.

2

Real Numbers

In mathematics, a real number is a value of a continuous quantity that can represent a distance along a line (or alternatively, a quantity that can be represented as an infinite decimal expansion). The adjective real in this context was introduced in the 17th century by René Descartes, who distinguished between real and imaginary roots of polynomials. The real numbers include all the rational numbers, such as the integer 5 and the fraction 4/3, and all the irrational numbers, such as $\sqrt{2}$ (1.41421356..., the square root of 2, an irrational algebraic number). Included within the irrationals are the real transcendental numbers, such as π (3.14159265...). In addition to measuring distance, real numbers can be used to measure quantities such as time, mass, energy, velocity, and many more. The set of real numbers is denoted using the symbol R or $\mathbb{R}$ and is sometimes called "the reals".

Real numbers can be thought of as points on an infinitely long line called the number line or real line, where the points corresponding to integers are equally spaced. Any real number can be determined by a possibly infinite decimal representation, such as that of 8.632, where each consecutive digit is measured in units one-tenth the size of the previous one. The real line can be thought of as a part of the complex plane, and the real numbers can be thought of as a part of the complex numbers.

These descriptions of the real numbers are not sufficiently rigorous by the modern standards of pure mathematics. The discovery of a suitably rigorous definition of the real numbers—indeed, the realization that a better definition was needed—was one of the most important developments of 19th-century mathematics. The current standard axiomatic definition is that real numbers form the unique Dedekind-complete ordered field (; + ; · ; <), up to an isomorphism,[a] whereas popular constructive definitions of real numbers include declaring them as equivalence classes of Cauchy sequences (of rational numbers), Dedekind cuts, or infinite decimal representations, together with precise interpretations for the arithmetic operations and the order relation. All these definitions satisfy the axiomatic definition and are thus equivalent.

The set of all real numbers is uncountable, in the sense that while both the set of all natural numbers and the set of all real numbers are infinite sets, there can be no one-to-one function from the real numbers to the natural numbers. In fact, the cardinality of the set of all real numbers, denoted by c and called the cardinality of the continuum, is strictly greater than the cardinality of the set of all natural numbers (denoted $5!_0$, 'aleph-naught').

The statement that there is no subset of the reals with cardinality strictly greater than $5!_0$ and strictly smaller than c is known as the continuum hypothesis (CH). It is neither provable nor refutable using the axioms of Zermelo–Fraenkel set theory including the axiom of choice (ZFC)—the standard foundation of modern mathematics. In fact, some models of ZFC satisfy CH, while others violate it.

Simple fractions were used by the Egyptians around 1000 BC; the Vedic "Shulba Sutras" ("The rules of chords") in c. 600 BC include what may be the first "use" of irrational numbers. The concept of irrationality was implicitly accepted by early Indian mathematicians such as Manava (c. 750–690 BC), who were aware that the square roots of certain numbers, such as 2 and 61, could not be exactly determined. Around 500 BC, the Greek mathematicians led by Pythagoras realized the need for irrational numbers, in particular the irrationality of the square root of 2.

The Middle Ages brought about the acceptance of zero, negative numbers, integers, and fractional numbers, first by Indian and Chinese mathematicians, and then by Arabic mathematicians, who were also the first to treat irrational numbers as algebraic objects (the latter being made possible by the development of algebra). Arabic mathematicians merged the concepts of "number" and "magnitude" into a more general idea of real numbers. The Egyptian mathematician Abû Kâmil Shujâ ibn Aslam (c. 850–930) was the first to accept irrational numbers as solutions to quadratic equations, or as coefficients in an equation (often in the form of square roots, cube roots and fourth roots).

In the 16th century, Simon Stevin created the basis for modern decimal notation, and insisted that there is no difference between rational and irrational numbers in this regard.

In the 17th century, Descartes introduced the term "real" to describe roots of a polynomial, distinguishing them from "imaginary" ones.

In the 18th and 19th centuries, there was much work on irrational and transcendental numbers. Johann Heinrich Lambert (1761) gave the first flawed proof that π cannot be rational; Adrien-Marie Legendre (1794) completed the proof, and showed that π is not the square root of a rational number. Paolo Ruffini (1799) and Niels Henrik Abel (1842) both constructed proofs of the Abel–Ruffini theorem: that the general quintic or higher equations cannot be solved by a general formula involving only arithmetical operations and roots.

Évariste Galois (1832) developed techniques for determining whether a given equation could be solved by radicals, which gave rise to the field of Galois theory. Joseph Liouville (1840) showed that neither e nor e^2 can be a root of an integer quadratic equation, and then established the existence of transcendental numbers; Georg Cantor (1873) extended and greatly simplified this proof. Charles Hermite (1873) first proved that e is transcendental, and Ferdinand von Lindemann (1882), showed that π is transcendental. Lindemann's proof was much simplified by Weierstrass (1885), still further by David Hilbert (1893), and has finally been made elementary by Adolf Hurwitz and Paul Gordan.

The development of calculus in the 18th century used the entire set of real numbers without having defined them rigorously. The first rigorous definition was published by Georg Cantor in 1871. In 1874, he showed that the set of all real numbers is uncountably infinite, but the set of all algebraic numbers is countably infinite. Contrary to widely held beliefs, his first method was not his famous diagonal argument, which he published in 1891. For more, see Cantor's first uncountability proof.

DEFINITION

The real number system can be defined axiomatically up to an isomorphism, which is described hereafter. There are also many ways to construct "the" real number system, and a popular approach involves starting from natural numbers, then defining rational numbers algebraically, and finally defining real numbers as equivalence classes of their Cauchy sequences or as Dedekind cuts, which are certain subsets of rational numbers. Another approach is to start from some rigorous axiomatization of Euclidean geometry (say of Hilbert or of Tarski), and then define the real number system geometrically. All these constructions of the real numbers have been shown to be equivalent, in the sense that the resulting number systems are isomorphic.

Axiomatic approach

Let R denote the set of all real numbers, then:

- The set R is a field, meaning that addition and multiplication are defined and have the usual properties.
- The field R is ordered, meaning that there is a total order e″ such that for all real numbers x, y and z:
  - o if x e″ y, then x + z e″ y + z;
  - o if x e″ 0 and y e″ 0, then xy e″ 0.
- The order is Dedekind-complete, meaning that every non-empty subset S of R with an upper bound in R has a least upper bound (a.k.a., supremum) in R.

The last property is what differentiates the real numbers from the rational numbers (and from other more exotic ordered fields). For example, {x ″ Q : x 2 < 2 } has a rational upper bound (e.g., 1.42), but no least rational upper bound, because is not rational.

These properties imply the Archimedean property (which is not implied by other definitions of completeness), which states that the set of integers has no upper bound in the reals. In fact, if this were false, then the integers would have a least upper bound N; then, N – 1 would not be an upper bound, and there would be an integer n such that n > N – 1, and thus n + 1 > N, which is a contradiction with the upper-bound property of N.

The real numbers are uniquely specified by the above properties. More precisely, given any two Dedekind-complete ordered fields R_1 and R_2, there exists a unique field isomorphism from R_1 to R_2. This uniqueness allows us to think of them as essentially the same mathematical object.

For another axiomatization of R, see Tarski's axiomatization of the reals.

Construction from the rational numbers

The real numbers can be constructed as a completion of the rational numbers, in such a way that a sequence defined by a decimal or binary expansion like (3; 3.1; 3.14; 3.141; 3.1415; ...) converges to a unique real number—in this case π. For details and other constructions of real numbers, see construction of the real numbers.

Properties

Basic properties

- Any non-zero real number is either negative or positive.
- The sum and the product of two non-negative real numbers is again a non-negative real number, i.e., they are closed under these operations, and form a positive cone, thereby giving rise to a linear order of the real numbers along a number line.

- The real numbers make up an infinite set of numbers that cannot be injectively mapped to the infinite set of natural numbers, i.e., there are uncountably infinitely many real numbers, whereas the natural numbers are called countably infinite. This establishes that in some sense, there are more real numbers than there are elements in any countable set.
- There is a hierarchy of countably infinite subsets of the real numbers, e.g., the integers, the rational numbers, the algebraic numbers and the computable numbers, each set being a proper subset of the next in the sequence. The complements of all these sets (irrational, transcendental, and non-computable real numbers) in the reals are all uncountably infinite sets.
- Real numbers can be used to express measurements of continuous quantities. They may be expressed by decimal representations, most of them having an infinite sequence of digits to the right of the decimal point; these are often represented like 324.823122147..., where the ellipsis (three dots) indicates that there would still be more digits to come. This hints at the fact that we can precisely denote only a few, selected real numbers with finitely many symbols.

More formally, the real numbers have the two basic properties of being an ordered field, and having the least upper bound property. The first says that real numbers comprise a field, with addition and multiplication as well as division by non-zero numbers, which can be totally ordered on a number line in a way compatible with addition and multiplication. The second says that, if a non-empty set of real numbers has an upper bound, then it has a real least upper bound. The second condition distinguishes the real numbers from the rational numbers: for example, the set of rational numbers whose square is less than 2 is a set with an upper bound (e.g. 1.5) but no (rational) least upper bound: hence the rational numbers do not satisfy the least upper bound property.

COMPLETENESS

A main reason for using real numbers is so that many sequences have limits. More formally, the reals are complete (in the sense of metric spaces or uniform spaces, which is a different sense than the Dedekind completeness of the order in the previous section):

A sequence (x_n) of real numbers is called a Cauchy sequence if for any å > 0 there exists an integer N (possibly depending on å) such that the distance |x_n ″ x_m| is less than å for all n and m that are both greater than N. This definition, originally provided by Cauchy, formalizes the fact that the x_n eventually come and remain arbitrarily close to each other.

A sequence (x_n) converges to the limit x if its elements eventually come and remain arbitrarily close to x, that is, if for any å > 0 there exists an integer N (possibly depending on å) such that the distance | x_n “ x | is less than å for n greater than N. Every convergent sequence is a Cauchy sequence, and the converse is true for real numbers, and this means that the topological space of the real numbers is complete.

The set of rational numbers is not complete. For example, the sequence (1; 1.4; 1.41; 1.414; 1.4142; 1.41421; ...), where each term adds a digit of the decimal expansion of the positive square root of 2, is Cauchy but it does not converge to a rational number (in the real numbers, in contrast, it converges to the positive square root of 2).

The completeness property of the reals is the basis on which calculus, and, more generally mathematical analysis are built. In particular, the test that a sequence is a Cauchy sequence allows proving that a sequence has a limit, without computing it, and even without knowing it.

For example, the standard series of the exponential function

$$e^x = \sum_{n=0}^{\infty} \frac{x^n}{n!}$$

converges to a real number for every x, because the sums

$$\sum_{n=N}^{M} \frac{x^n}{n!}$$

can be made arbitrarily small (independently of M) by choosing N sufficiently large. This proves that the sequence is Cauchy, and thus converges, showing that e^x is well defined for every x.

Completeness of the real numbers

Intuitively, completeness implies that there are not any "gaps" (in Dedekind's terminology) or "missing points" in the real number line. This contrasts with the rational numbers, whose corresponding number line has a "gap" at each irrational value. In the decimal number system, completeness is equivalent to the statement that any infinite string of decimal digits is actually a decimal representation for some real number.

Depending on the construction of the real numbers used, completeness may take the form of an axiom (the completeness axiom), or may be a theorem proven from the construction. There are many equivalent forms of completeness, the most prominent being Dedekind completeness and Cauchy completeness (completeness as a metric space).

Forms of completeness

The real numbers can be defined synthetically as an ordered field satisfying some version of the completeness axiom. Different versions of this axiom are all equivalent in the sense that any ordered field that satisfies one form of completeness satisfies all of them, apart from Cauchy completeness and nested intervals theorem, which are strictly weaker in that there are non Archimedean fields that are ordered and Cauchy complete. When the real numbers are instead constructed using a model, completeness becomes a theorem or collection of theorems.

Least upper bound property

The least-upper-bound property states that every nonempty subset of real numbers having an upper bound must have a least upper bound (or supremum) in the set of real numbers.

The rational number line Q does not have the least upper bound property. An example is the subset of rational numbers

$$S = \{x \in \mathbb{Q}^* x^2 < 2\}.$$

This set has an upper bound. However, this set has no least upper bound in Q: the least upper bound as a subset of the reals would be "2, but it does not exist in Q. For any upper bound x " Q, there is another upper bound y " Q with y < x.

For instance, take x = 1.5, then x is certainly an upper bound of S, since x is positive and x^2 = 2.25 e" 2; that is, no element of S is larger than x. However, we can choose a smaller upper bound, say y = 1.45; this is also an upper bound of S for the same reasons, but it is smaller than x, so x is not a least-upper-bound of S. We can proceed similarly to find an upper bound of S that is smaller than y, say z = 1.42, etc., such that we never find a least-upper-bound of S in Q.

The least upper bound property can be generalized to the setting of partially ordered sets. See completeness (order theory).

Dedekind completeness

Dedekind completeness is the property that every Dedekind cut of the real numbers is generated by a real number. In a synthetic approach to the real numbers, this is the version of completeness that is most often included as an axiom.

The rational number line Q is not Dedekind complete. An example is the Dedekind cut

$$L = \{x \in \mathbb{Q}^* x^2 \leq 2 \vee x < 0\}.$$

$R = \{x \in \mathbb{Q}^* x^2 \geq 2 \wedge x > 0\}$.

L does not have a maximum and R does not have a minimum, so this cut is not generated by a rational number.

There is a construction of the real numbers based on the idea of using Dedekind cuts of rational numbers to name real numbers; e.g. the cut (L,R) described above would name $\sqrt{2}$. If one were to repeat the construction of real numbers with Dedekind cuts (i.e., "close" the set of real numbers by adding all possible Dedekind cuts), one would obtain no additional numbers because the real numbers are already Dedekind complete.

Cauchy completeness

Cauchy completeness is the statement that every Cauchy sequence of real numbers converges.

The rational number line Q is not Cauchy complete. An example is the following sequence of rational numbers:

3, 3.1, 3.14, 3.142, 3.1416, ...

Here the nth term in the sequence is the nth decimal approximation for pi. Though this is a Cauchy sequence of rational numbers, it does not converge to any rational number. (In this real number line, this sequence converges to pi.)

Cauchy completeness is related to the construction of the real numbers using Cauchy sequences. Essentially, this method defines a real number to be the limit of a Cauchy sequence of rational numbers.

In mathematical analysis, Cauchy completeness can be generalized to a notion of completeness for any metric space. See complete metric space.

For an ordered field, Cauchy completeness is weaker than the other forms of completeness on this page. But Cauchy completeness and the Archimedean property taken together are equivalent to the others.

Nested intervals theorem

The nested interval theorem is another form of completeness. Let I_n = $[a_n, b_n]$ be a sequence of closed intervals, and suppose that these intervals are nested in the sense that

$$I_1 \supset I_2 \supset I_3 \supset \cdots$$

Moreover, assume that b_n " a_n '! 0 as n '! +". The nested interval theorem states that the intersection of all of the intervals I_n contains exactly one point.

The rational number line does not satisfy the nested interval theorem. For example, the sequence (whose terms are derived from the digits of pi in the suggested way)

$$[3,4] \supset [3.1,3.2] \supset [3.14,3.15] \supset [3.141,3.142] \supset \cdots$$

is a nested sequence of closed intervals in the rational numbers whose intersection is empty. (In the real numbers, the intersection of these intervals contains the number pi.)

Nested intervals theorem shares the same logical status as Cauchy completeness in this spectrum of expressions of completeness. In other words, nested intervals theorem by itself is weaker than other forms of completeness, although taken together with Archimedean property, it is equivalent to the others.

Monotone convergence theorem

The monotone convergence theorem (described as the fundamental axiom of analysis by Körner states that every nondecreasing, bounded sequence of real numbers converges. This can be viewed as a special case of the least upper bound property, but it can also be used fairly directly to prove the Cauchy completeness of the real numbers.

Bolzano–Weierstrass theorem

The Bolzano–Weierstrass theorem states that every bounded sequence of real numbers has a convergent subsequence. Again, this theorem is equivalent to the other forms of completeness given above.

THE INTERMEDIATE VALUE THEOREM

The intermediate value theorem states that every continuous function that attains both negative and positive values has a root. This is a consequence of the least upper bound property, but it can also be used to prove the least upper bound property if treated as an axiom. (The definition of continuity does not depend on any form of completeness, so this is not circular.)

"The complete ordered field"

The real numbers are often described as "the complete ordered field", a phrase that can be interpreted in several ways.

First, an order can be lattice-complete. It is easy to see that no ordered field can be lattice-complete, because it can have no largest element (given any element z, z + 1 is larger).

Additionally, an order can be Dedekind-complete, see § Axiomatic approach. The uniqueness result at the end of that section justifies using the word "the" in the phrase "complete ordered field" when this is the sense of "complete" that is meant. This sense of completeness is most closely related to the construction of the reals from Dedekind cuts, since that construction starts from an ordered field (the rationals) and then forms the Dedekind-completion of it in a standard way.

These two notions of completeness ignore the field structure. However, an ordered group (in this case, the additive group of the field) defines a uniform structure, and uniform structures have a notion of completeness; the description in § Completeness is a special case. (We refer to the notion of completeness in uniform spaces rather than the related and better known notion for metric spaces, since the definition of metric space relies on already having a characterization of the real numbers.) It is not true that R is the only uniformly complete ordered field, but it is the only uniformly complete Archimedean field, and indeed one often hears the phrase "complete Archimedean field" instead of "complete ordered field". Every uniformly complete Archimedean field must also be Dedekind-complete (and vice versa), justifying using "the" in the phrase "the complete Archimedean field". This sense of completeness is most closely related to the construction of the reals from Cauchy sequences (the construction carried out in full in this article), since it starts with an Archimedean field (the rationals) and forms the uniform completion of it in a standard way.

But the original use of the phrase "complete Archimedean field" was by David Hilbert, who meant still something else by it. He meant that the real numbers form the largest Archimedean field in the sense that every other Archimedean field is a subfield of R. Thus R is "complete" in the sense that nothing further can be added to it without making it no longer an Archimedean field. This sense of completeness is most closely related to the construction of the reals from surreal numbers, since that construction starts with a proper class that contains every ordered field (the surreals) and then selects from it the largest Archimedean subfield.

Advanced properties

The reals are uncountable; that is, there are strictly more real numbers than natural numbers, even though both sets are infinite. In fact, the cardinality of the reals equals that of the set of subsets (i.e. the power set) of the natural numbers, and Cantor's diagonal argument states that the latter set's cardinality is strictly greater than the cardinality of N. Since the set of algebraic numbers is countable, almost all real numbers are transcendental. The non-existence of a subset of the reals with cardinality strictly between that of the integers and the reals is known as the continuum hypothesis. The continuum hypothesis can neither be proved nor be disproved; it is independent from the axioms of set theory.

As a topological space, the real numbers are separable. This is because the set of rationals, which is countable, is dense in the real numbers. The irrational numbers are also dense in the real numbers, however they are uncountable and have the same cardinality as the reals.

The real numbers form a metric space: the distance between x and y is defined as the absolute value | x " y |. By virtue of being a totally ordered set, they also carry an order topology; the topology arising from the metric and the one arising from the order are identical, but yield different presentations for the topology—in the order topology as ordered intervals, in the metric topology as epsilon-balls. The Dedekind cuts construction uses the order topology presentation, while the Cauchy sequences construction uses the metric topology presentation. The reals form a contractible (hence connected and simply connected), separable and complete metric space of Hausdorff dimension 1. The real numbers are locally compact but not compact. There are various properties that uniquely specify them; for instance, all unbounded, connected, and separable order topologies are necessarily homeomorphic to the reals.

Every nonnegative real number has a square root in R, although no negative number does. This shows that the order on R is determined by its algebraic structure. Also, every polynomial of odd degree admits at least one real root: these two properties make R the premier example of a real closed field. Proving this is the first half of one proof of the fundamental theorem of algebra.

The reals carry a canonical measure, the Lebesgue measure, which is the Haar measure on their structure as a topological group normalized such that the unit interval [0;1] has measure 1. There exist sets of real numbers that are not Lebesgue measurable, e.g. Vitali sets.

The supremum axiom of the reals refers to subsets of the reals and is therefore a second-order logical statement. It is not possible to characterize the reals with first-order logic alone: the Löwenheim–Skolem theorem implies that there exists a countable dense subset of the real numbers satisfying exactly the same sentences in first-order logic as the real numbers themselves. The set of hyperreal numbers satisfies the same first order sentences as R. Ordered fields that satisfy the same first-order sentences as R are called nonstandard models of R. This is what makes nonstandard analysis work; by proving a first-order statement in some nonstandard model (which may be easier than proving it in R, we know that the same statement must also be true of R.

The field R of real numbers is an extension field of the field Q of rational numbers, and R can therefore be seen as a vector space over Q. Zermelo–Fraenkel set theory with the axiom of choice guarantees the existence of a basis of this vector space: there exists a set B of real numbers such that every real number can be written uniquely as a finite linear combination of elements of this set, using rational coefficients only, and such that no element of B is a rational linear combination of the others. However, this existence theorem is purely theoretical, as such a base has never been explicitly described.

The well-ordering theorem implies that the real numbers can be well-ordered if the axiom of choice is assumed: there exists a total order on R with the property that every non-empty subset of R has a least element in this ordering. (The standard ordering d″ of the real numbers is not a well-ordering since e.g. an open interval does not contain a least element in this ordering.) Again, the existence of such a well-ordering is purely theoretical, as it has not been explicitly described. If V=L is assumed in addition to the axioms of ZF, a well ordering of the real numbers can be shown to be explicitly definable by a formula.

A real number may be either computable or uncomputable; either algorithmically random or not; and either arithmetically random or not.

Applications and connections to other areas

REAL NUMBERS AND LOGIC

The real numbers are most often formalized using the Zermelo–Fraenkel axiomatization of set theory, but some mathematicians study the real numbers with other logical foundations of mathematics. In particular, the real numbers are also studied in reverse mathematics and in constructive mathematics.

The hyperreal numbers as developed by Edwin Hewitt, Abraham Robinson and others extend the set of the real numbers by introducing infinitesimal and infinite numbers, allowing for building infinitesimal calculus in a way closer to the original intuitions of Leibniz, Euler, Cauchy and others.

Edward Nelson's internal set theory enriches the Zermelo–Fraenkel set theory syntactically by introducing a unary predicate "standard". In this approach, infinitesimals are (non-"standard") elements of the set of the real numbers (rather than being elements of an extension thereof, as in Robinson's theory).

The continuum hypothesis posits that the cardinality of the set of the real numbers is $5!_1$; i.e. the smallest infinite cardinal number after $5!_0$, the cardinality of the integers. Paul Cohen proved in 1963 that it is an axiom independent of the other axioms of set theory; that is: one may choose either the continuum hypothesis or its negation as an axiom of set theory, without contradiction.

In physics

In the physical sciences, most physical constants such as the universal gravitational constant, and physical variables, such as position, mass, speed, and electric charge, are modeled using real numbers. In fact, the fundamental physical theories such as classical mechanics, electromagnetism, quantum mechanics, general relativity and the standard model are described using mathematical structures, typically smooth manifolds or Hilbert spaces, that are based on the real numbers, although actual measurements of physical quantities are of finite accuracy and precision.

Physicists have occasionally suggested that a more fundamental theory would replace the real numbers with quantities that do not form a continuum, but such proposals remain speculative.

In computation

With some exceptions, most calculators do not operate on real numbers. Instead, they work with finite-precision approximations called floating-point numbers. In fact, most scientific computation uses floating-point arithmetic. Real numbers satisfy the usual rules of arithmetic, but floating-point numbers do not.

Computers cannot directly store arbitrary real numbers with infinitely many digits. The achievable precision is limited by the number of bits allocated to store a number, whether as floating-point numbers or arbitrary-precision numbers. However, computer algebra systems can operate on irrational quantities exactly by manipulating formulas for them (such as 2, arcsin (2/23), or $\int_0^1 x^x\,dx$) rather than their rational or decimal approximation. It is not in general possible to determine whether two such expressions are equal (the constant problem).

A real number is called computable if there exists an algorithm that yields its digits. Because there are only countably many algorithms, but an uncountable number of reals, almost all real numbers fail to be computable. Moreover, the equality of two computable numbers is an undecidable problem. Some constructivists accept the existence of only those reals that are computable. The set of definable numbers is broader, but still only countable.

"Reals" in set theory

In set theory, specifically descriptive set theory, the Baire space is used as a surrogate for the real numbers since the latter have some topological properties (connectedness) that are a technical inconvenience. Elements of Baire space are referred to as "reals".

MULTIPLICATION

Properties of Multiplying Real Numbers

There are five properties of multiplication that are important for you to know:

Commutative Property

The commutative property of multiplication states that the order in which two numbers are multiplied does not affect the sum. If a and b are real numbers, then a×b=b×a.

Closure Property

The product of any two real numbers will result in a real number. This is known as the closure property of multiplication. In general, the closure property states that the product of any two real numbers is a unique real number. If a,b and c are real numbers, then a×b=c.

Associative Property

The order in which three or more real numbers are grouped for multiplication will not affect the product. This is known as the associative property of multiplication. The result will always be the same real number. In general, the associative property states that the order in which the numbers are grouped for multiplication does not change the product. If a,b and c are real numbers, then (a×b)×c=a×(b×c).

Multiplicative Identity

When any real number is multiplied by the number one, the real number does not change. This is true whether the real number is positive or negative. The number 1 is called the multiplicative identity or the identity element of multiplication. The product of a number and one is the number. This is called the identity property of multiplication. If a is a real number, then a×1=a.

Multiplicative Inverse

If a is a nonzero real number, then the reciprocal or multiplicative inverse of a is 1a. The product of any nonzero real number and its reciprocal is always one. The number 1 is called the multiplicative identity or the identity element of multiplication. Therefore, the product of a and its reciprocal is the identity element of multiplication (one). This is known as the inverse property of multiplication. If a is a nonzero real number, then a×1a=1.

Now, let's apply the properties to the following problems:

1. Does ("3)×(+2)=(+2)×("3)?

(“3)×(+2)=(+2)×(“3)=”6

This is an example of the commutative property of multiplication.

2. Does (“6)×(+3) equal a real number?

(“6)×(+3)=”18, a real number. This is an example of the closure property of multiplication.

3. Does (“3×2)×2=”3×(2×2)?

(“3×2)×2=”3×(2×2)=”12 Even though the numbers are grouped differently, the result is the same. This is an example of the associative property of multiplication.

4. Does 8×1=8?

Yes. This is an example of the identity property of multiplication.

5. Does 7×17=1?

Yes. This is an example of the inverse property of multiplication.

Examples

Example 1.6.9.1

Earlier, you were given a problem that asked if the statement (“2)×(“3)=(“3)×(“2)=6 is true.

Solution

The order in which you multiplied the numbers did not affect the answer. This is an example of the commutative property of multiplication.

Example 1.6.9.1

Multiply using the properties of multiplication: (6×1/6)×(3×”1)

Solution

(6×1/6)×(3×”1)=6/6×”3=1×”3=”3

Example 1.6.9.1

What property of multiplication justifies the statement (“9×5)×2=”9×(5×2)?

Solution

The associative property of multiplication because the numbers were regrouped with parentheses.

Example 1.6.9.1

What property of multiplication justifies the statement “176×1=”176?

Solution

The identity property of multiplication because any number multiplied by 1 is itself.

ADDITION

When adding real numbers with the same sign the sum will have the same sign as the numbers added.

3+2=5

″7+(″2)=″9

When adding real numbers with different signs you subtract the lesser absolute value from greater one. The sum will then have the same sign as the number with the greater absolute value.

″7+2=″5

There are a couple of properties of addition:

The additive commutative property tells us that the order in which you add the numbers does not change the sum.

a+b=b+a

And the additive associative property tells us that it also that the order in which we group three or more numbers does not affect the sum.

(a+b)+c=a+(b+c)

The additive identity property tells us that the sum of a number and 0 is always the number.

a+0=a

And the additive inverse property tells us that if you add a number with its opposite you will always get 0

a+(″a)=0

SUBTRACTION

You know by now that

10″2=10+(″2)=8

This means that subtracting 2 from 10 is the same as adding negative 2 to 10. This is called the subtraction rule.

a″b=a+(″b)

REAL NUMBERS: POSITIVITY

In mathematics, the set of positive real numbers, R > 0 = { x ″ R #″ x > 0 }, is the subset of those real numbers that are greater than zero. The non-negative real numbers, R e″ 0 = { x ″ R #″ x e″ 0 }, also include zero. Although the symbols R_+ and R^+ are ambiguously used for either of these, the notation R_+ or R^+ for {x ″ R #″ x e″ 0} and $R^+_{″}$ or $R_{″}^+$ for { x ″ R #″ x > 0 } has also been widely employed, is aligned with the practice in algebra of denoting the exclusion of the zero element with a star, and should be understandable to most practicing mathematicians.

In a complex plane, R > 0 is identified with the positive real axis, and is usually drawn as a horizontal ray. This ray is used as reference in the polar form of a complex number. The real positive axis corresponds to complex numbers z = | z | e i ö, with argument ö = 0.

Ratio scale

Among the levels of measurement the ratio scale provides the finest detail. The division function takes a value of one (1) when numerator and denominator are equal. Other ratios are compared to one by logarithms, often common logarithm using base 10. The ratio scale then segments by orders of magnitude used in science and technology, expressed in various units of measurement.

An early expression of ratio scale was articulated geometrically by Eudoxus: "it was ... in geometrical language that the general theory of proportion of Eudoxus was developed, which is equivalent to a theory of positive real numbers."

Generalizations and extensions

The real numbers can be generalized and extended in several different directions:

- The complex numbers contain solutions to all polynomial equations and hence are an algebraically closed field unlike the real numbers. However, the complex numbers are not an ordered field.
- The affinely extended real number system adds two elements +" and ". It is a compact space. It is no longer a field, or even an additive group, but it still has a total order; moreover, it is a complete lattice.
- The real projective line adds only one value ". It is also a compact space. Again, it is no longer a field, or even an additive group. However, it allows division of a non-zero element by zero. It has cyclic order described by a separation relation.
- The long real line pastes together $5!_1$* + $5!_1$ copies of the real line plus a single point (here $5!_1$* denotes the reversed ordering of $5!_1$) to create an ordered set that is "locally" identical to the real numbers, but somehow longer; for instance, there is an order-preserving embedding of $5!_1$ in the long real line but not in the real numbers. The long real line is the largest ordered set that is complete and locally Archimedean. As with the previous two examples, this set is no longer a field or additive group.
- Ordered fields extending the reals are the hyperreal numbers and the surreal numbers; both of them contain infinitesimal and infinitely large numbers and are therefore non-Archimedean ordered fields.
- Self-adjoint operators on a Hilbert space (for example, self-adjoint square complex matrices) generalize the reals in many respects: they can be

ordered (though not totally ordered), they are complete, all their eigenvalues are real and they form a real associative algebra. Positive-definite operators correspond to the positive reals and normal operators correspond to the complex numbers.

Square roots and real numbers

In an earlier chapter we learned that

3^2=3Å"3=9

We said that 9 was the square of 3. The square of -3 is 9 as well

("3)^2=("3)Å"("3)=9

3 and -3 are said to be the square roots of 9.

All positive real numbers has two square roots, one positive square root and one negative square root. The positive square root is sometimes referred to as the principal square root. The reason that we have two square roots is exemplified above. The product of two numbers is positive if both numbers have the same sign as is the case with squares and square roots

$$a^2 = a \cdot a = (-a) \cdot (-a)$$

A square root is written with a radical symbol " and the number or expression inside the radical symbol, below denoted a, is called the radicand.

$$\sqrt{a}$$

To indicate that we want both the positive and the negative square root of a radicand we put the symbol ± (read as plus minus) in front of the root.

$$\pm\sqrt{9} = \pm 3$$

Zero has one square root which is 0.

$$\sqrt{0} = 0$$

Negative numbers don't have real square roots since a square is either positive or 0.

If the square root of an integer is another integer then the square is called a perfect square. For example 25 is a perfect square since

$$\pm\sqrt{25} = \pm 5$$

If the radicand is not a perfect square i.e. the square root is not a whole number than you have to approximate the square root

$$\pm\sqrt{3} = \pm 1.73205... \approx \pm 1.7$$

The square roots of numbers that are not a perfect square are members of the irrational numbers. This means that they can't be written as the quotient of two integers. The decimal form of an irrational number will neither terminate nor repeat. The irrational numbers together with the rational numbers constitutes the real numbers.

Example

irrational number $\Rightarrow \sqrt{19} \approx 4.35889...$

rationalnumber$\Rightarrow$!$0.5=\frac{1}{2}$

INEQUALITIES

In mathematics, an inequality is a relation which makes a non-equal comparison between two numbers or other mathematical expressions. It is used most often to compare two numbers on the number line by their size. There are several different notations used to represent different kinds of inequalities:

- The notation a < b means that a is less than b.
- The notation a > b means that a is greater than b.

In either case, a is not equal to b. These relations are known as strict inequalities, meaning that a is strictly less than or strictly greater than b. Equivalence is excluded.

In contrast to strict inequalities, there are two types of inequality relations that are not strict:

- The notation a d″ b or a }* b means that a is less than or equal to b (or, equivalently, at most b, or not greater than b).
- The notation a e″ b or a ~* b means that a is greater than or equal to b (or, equivalently, at least b, or not less than b).

The relation not greater than can also be represented by a o″ b, the symbol for "greater than" bisected by a slash, "not". The same is true for not less than and a n″ b.

The notation a ‴ b means that a is not equal to b; this inequation sometimes is considered a form of strict inequality. It does not say that one is greater than the other; it does not even require a and b to be member of an ordered set.

In engineering sciences, less formal use of the notation is to state that one quantity is "much greater" than another, normally by several orders of magnitude.

- The notation a j″ b means that a is much less than b.
- The notation a k″ b means that a is much greater than b.

This implies that the lesser value can be neglected with little effect on the accuracy of an approximation (such as the case of ultrarelativistic limit in physics).

In all of the cases above, any two symbols mirroring each other are symmetrical; a < b and b > a are equivalent, etc.

Properties on the number line

Inequalities are governed by the following properties. All of these properties also hold if all of the non-strict inequalities (d and e) are replaced by their corresponding strict inequalities (< and >) and — in the case of applying a function — monotonic functions are limited to strictly monotonic functions.

Converse

The relations d and e are each other's converse, meaning that for any real numbers a and b:

a d b and b e a are equivalent.

Transitivity

The transitive property of inequality states that for any real numbers a, b, c:

If a d b and b d c, then a d c.

If either of the premises is a strict inequality, then the conclusion is a strict inequality:

If a d″ b and b < c, then a < c.

If a < b and b d″ c, then a < c.

Addition and subtraction

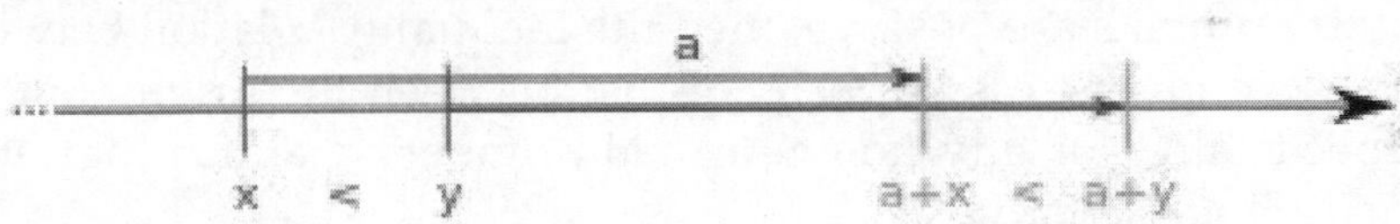

If x < y, then x + a < y + a.

A common constant c may be added to or subtracted from both sides of an inequality. So, for any real numbers a, b, c:

If a d″ b, then a + c d″ b + c and a “ c d″ b “ c.

In other words, the inequality relation is preserved under addition (or subtraction) and the real numbers are an ordered group under addition.

Multiplication and division

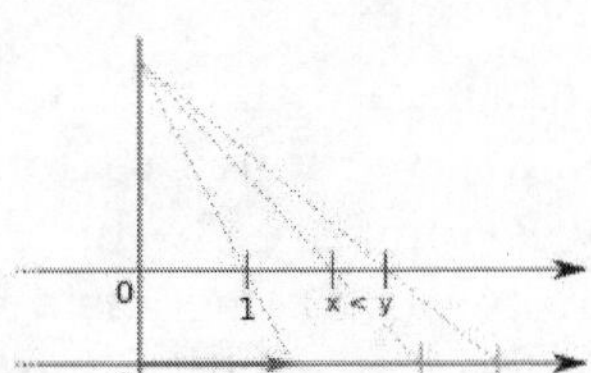

If x < y and a > 0, then ax < ay.

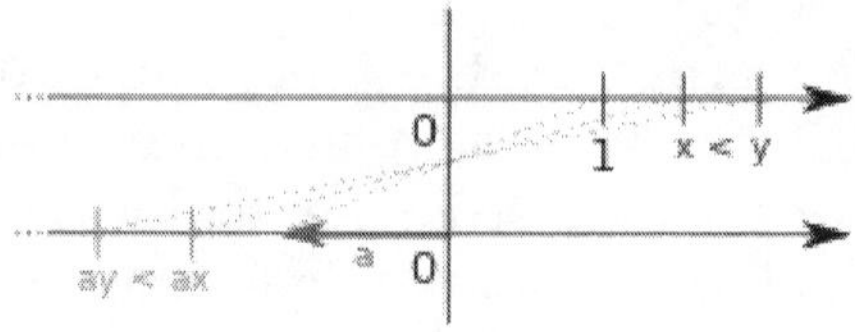

If x < y and a < 0, then ax > ay.

The properties that deal with multiplication and division state that for any real numbers, a, b and non-zero c:

If a d b and c > 0, then ac d bc and a/c d b/c.

If a d b and c < 0, then ac e bc and a/c e b/c.

In other words, the inequality relation is preserved under multiplication and division with positive constant, but is reversed when a negative constant is involved. More generally, this applies for an ordered field. For more information, see § Ordered fields.

Additive inverse

The property for the additive inverse states that for any real numbers a and b:

If a d b, then a e b.

Multiplicative inverse

If both numbers are positive, then the inequality relation between the multiplicative inverses is opposite of that between the original numbers. More specifically, for any non-zero real numbers a and b that are both positive (or both negative):

If a d b, then 1/a e 1/b.

All of the cases for the signs of a and b can also be written in chained notation, as follows:

If 0 < a d b, then 1/a e 1/b > 0.

If a d b < 0, then 0 > 1/a e 1/b.

If a < 0 < b, then 1/a < 0 < 1/b.

Applying a function to both sides

Any monotonically increasing function, by its definition, may be applied to both sides of an inequality without breaking the inequality relation (provided that both expressions are in the domain of that function). However, applying a monotonically decreasing function to both sides of an inequality means the inequality relation would be reversed. The rules for the additive inverse, and the multiplicative inverse for positive numbers, are both examples of applying a monotonically decreasing function.

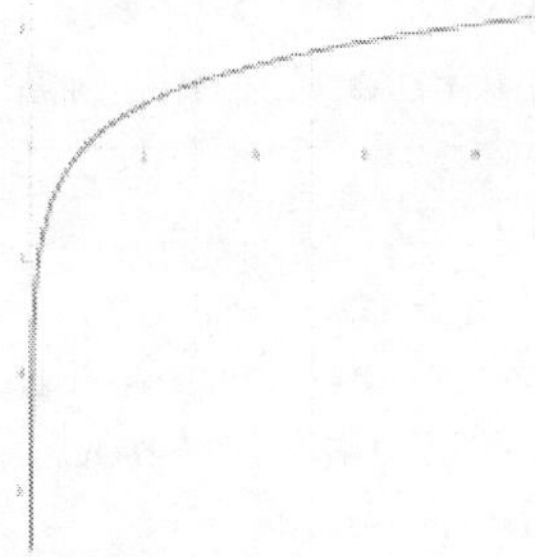

If the inequality is strict (a < b, a > b) and the function is strictly monotonic, then the inequality remains strict. If only one of these conditions is strict, then the resultant inequality is non-strict. In fact, the rules for additive and multiplicative inverses are both examples of applying a strictly monotonically decreasing function.

A few examples of this rule are:

- Raising both sides of an inequality to a power n > 0 (equiv., ″n < 0), when a and b are positive real numbers:
  0 d″ a d″ b Ô! 0 d″ a^n d″ b^n.
  0 d″ a d″ b Ô! $a^{''n}$ e″ $b^{''n}$ e″ 0.
- Taking the natural logarithm on both sides of an inequality, when a and b are positive real numbers:
  0 < a d″ b Ô! ln(a) d″ ln(b).
  0 < a < b Ô! ln(a) < ln(b).

(this is true because the natural logarithm is a strictly increasing function.)

Formal definitions and generalizations

A (non-strict) partial order is a binary relation d″ over a set P which is reflexive, antisymmetric, and transitive. That is, for all a, b, and c in P, it must satisfy the three following clauses:

1. a d″ a (reflexivity)
2. if a d″ b and b d″ a, then a = b (antisymmetry)
3. if a d″ b and b d″ c, then a d″ c (transitivity)

A set with a partial order is called a partially ordered set. Those are the very basic axioms that every kind of order has to satisfy. Other axioms that exist for other definitions of orders on a set P include:

1. For every a and b in P, a d″ b or b d″ a (total order).
2. For all a and b in P for which a < b, there is a c in P such that a < c < b (dense order).
3. Every non-empty subset of P with an upper bound has a least upper bound (supremum) in P (least-upper-bound property).

ORDERED FIELDS

If (F, +, ×) is a field and d″ is a total order on F, then (F, +, ×, d″) is called an ordered field if and only if:

- a d″ b implies a + c d″ b + c;
- 0 d″ a and 0 d″ b implies 0 d″ a × b.

Both (Q, +, ×, d″) and (R, +, ×, d″) are ordered fields, but d″ cannot be defined in order to make (C, +, ×, d″) an ordered field, because “1 is the square of i and would therefore be positive.

Besides from being an ordered field, R also has the Least-upper-bound property. In fact, R can be defined as the only ordered field with that quality.

Chained notation

The notation a < b < c stands for “a < b and b < c”, from which, by the transitivity property above, it also follows that a < c. By the above laws, one can add or subtract the same number to all three terms, or multiply or divide all three terms by same nonzero number and reverse all inequalities if that number is negative. Hence, for example, a < b + e < c is equivalent to a “ e < b < c “ e.

This notation can be generalized to any number of terms: for instance, a_1 d″ a_2 d″ ... d″ a_n means that a_i d″ a_{i+1} for i = 1, 2, ..., n “ 1. By transitivity, this condition is equivalent to a_i d″ a_j for any 1 d″ i d″ j d″ n.

When solving inequalities using chained notation, it is possible and sometimes necessary to evaluate the terms independently. For instance, to solve the inequality 4x < 2x + 1 d″ 3x + 2, it is not possible to isolate x in any one part of the inequality through addition or subtraction. Instead, the inequalities must be solved independently, yielding x < 1/2 and x e″ “1 respectively, which can be combined into the final solution “1 d″ x < 1/2.

Occasionally, chained notation is used with inequalities in different directions, in which case the meaning is the logical conjunction of the inequalities between adjacent terms. For example, the defining condiction of a zigzag poset is written as $a_1 < a_2 > a_3 < a_4 > a_5 < a_6 > \ldots$. Mixed chained notation is used more often with compatible relations, like <, =, d″. For instance, a < b = c d″ d means that a < b, b = c, and c d″ d. This notation exists in a few programming languages such as Python. In contrast, in programming languages that provide an ordering on the type of comparison results, such as C, even homogeneous chains may have a completely different meaning.

How to Use Inequalities

Inequalities can be used in many different ways. First, they can be used to show the relationship between two quantities. For example:

1 < 13

and

7.5 > 7.2

Inequalities are a good way to show the differences between real numbers that might not be easily apparent at a glance.

Real numbers are all numbers that are not imaginary. They include the following sets of numbers:

- Whole numbers, which are non-negative numbers that are not fractions or decimals.
- Integers , which are all numbers that are not fractions or decimals, including negative numbers.
- Rational numbers are positive and negative numbers that include fractions and decimals.
- Irrational numbers include decimal numbers that cannot be written as fractions, for example: e and pi.

Here's another example:

sqrt. 2 < pi

2/3 > 0.15

The second way you can use inequalities is to solve problems, such as:

Given the inequality a < b, write another inequality with the same meaning.

Inequality with all real numbers as solutions

An inequality with all real numbers as solutions is easy to solve or easy to recognize. See a few examples below.

Example #1

Solve x - x > -1

x - x > -1

x - x = 0, so we get 0 > -1

Since 0 is always bigger than -1, this inequality is always true. Therefore, all real numbers are solutions.

Example #2

Solve 5x + x + 3 > 6x + -4

5x + x + 3 > 6x + -4

Subtract 3 from each side of the inequality

5x + x + 3 - 3 > 6x + -4 - 3

5x + x + 0 > 6x + -4 + -3

5x + x > 6x + -7

Simplify by adding 5x and x

6x > 6x + -7

Subtract 6x from each side of the inequality

6x - 6x > 6x - 6x + -7

0 > 0 + -7

0 > -7

Since 0 is always bigger than -7, this inequality is always true. Therefore, all real numbers are solutions.

Compound inequality with all real numbers as solutions

Example #3

Solve 8x + 4 d″ 20 or 3x - 2 > 1

Solve 8x + 4 d″ 20

8x + 4 d″ 20

Subtract 4 from each side of the inequality

8x + 4 - 4 d″ 20 - 4

8x d″ 16

Divide each side by 8

8x/8 d″ 16/8

x d″ 2

Solve 3x - 2 > 1

3x - 2 > 1

Add 2 to each side of the inequality

3x - 2 + 2 > 1 + 2

3x - 2 + 2 > 1 + 2

3x > 3

Divide each side by 3

3x/3 > 3/3

x > 1

The solution is x d″ 2 or x > 1 and this includes all real numbers as shown in the graph below.

Absolute value inequality with all real numbers as solutions

Example #4

Solve |x| > -10

The absolute value of a number is always positive.

For example, suppose x is positive and choose x = 2

|2| = 2 and 2 is always bigger -10

Suppose x is negative and choose x = -2

$|-2| = 2$ and 2 is always bigger than -10

Therefore, $|x| > -10$ has all real numbers as solutions

Sharp inequalities

An inequality is said to be sharp if it cannot be relaxed and still be valid in general. Formally, a universally quantified inequality ö is called sharp if, for every valid universally quantified inequality ø, if ø ⇒! ö holds, then ø Ô! ö also holds. For instance, the inequality "a " "!. a^2 e" 0 is sharp, whereas the inequality "a " "!. a^2 e" "1 is not sharp.

Systems of inequalities

Systems of linear inequalities can be simplified by Fourier–Motzkin elimination. The cylindrical algebraic decomposition is an algorithm that allows testing whether a system of polynomial equations and inequalities has solutions, and, if solutions exist, describing them. The complexity of this algorithm is doubly exponential in the number of variables. It is an active research domain to design algorithms that are more efficient in specific cases.

3

Decimal and Percent

The decimal numeral system is the standard system for denoting integer and non-integer numbers. It is the extension to non-integer numbers of the Hindu–Arabic numeral system. The way of denoting numbers in the decimal system is often referred to as decimal notation.

A decimal numeral (also often just decimal or, less correctly, decimal number), refers generally to the notation of a number in the decimal numeral system. Decimals may sometimes be identified by a decimal separator (usually "." or "," as in 25.9703 or 3,1415). Decimal may also refer specifically to the digits after the decimal separator, such as in "3.14 is the approximation of π to two decimals". Zero-digits after a decimal separator serve the purpose of signifying the precision of a value.

The numbers that may be represented in the decimal system are the decimal fractions. That is, fractions of the form $a/10^n$, where a is an integer, and n is a non-negative integer.

The decimal system has been extended to infinite decimals for representing any real number, by using an infinite sequence of digits after the decimal separator. In this context, the decimal numerals with a finite number of non-zero digits after the decimal separator are sometimes called terminating decimals. A repeating decimal is an infinite decimal that, after some place, repeats indefinitely the same sequence of digits (e.g., 5.123144144144144... = 5.123144). An infinite decimal represents a rational number, the quotient of two integers, if and only if it is a repeating decimal or has a finite number of non-zero digits.

DECIMAL NOTATION

For writing numbers, the decimal system uses ten decimal digits, a decimal mark, and, for negative numbers, a minus sign "". The decimal digits are 0, 1, 2, 3, 4, 5, 6, 7, 8, 9; the decimal separator is the dot "." in many countries (mostly English-speaking), and a comma "," in other countries.

For representing a non-negative number, a decimal numeral consists of

- either a (finite) sequence of digits (such as "2017"), where the entire sequence represents an integer,

  $a_m a_{m-1} \dots a_0$

- or a decimal mark separating two sequences of digits (such as "20.70828")

  $a_m a_{m-1} \dots a_0 . b_1 b_2 \dots b_n$

If m > 0, that is, if the first sequence contains at least two digits, it is generally assumed that the first digit a_m is not zero. In some circumstances it may be useful to have one or more 0's on the left; this does not change the value represented by the decimal: for example, 3.14 = 03.14 = 003.14. Similarly, if the final digit on the right of the decimal mark is zero—that is, if $b_n = 0$—it may be removed; conversely, trailing zeros may be added after the decimal mark without changing the represented number; [note 1] for example, 15 = 15.0 = 15.00 and 5.2 = 5.20 = 5.200.

For representing a negative number, a minus sign is placed before a_m.

The numeral $a_m a_{m-1} \dots a_0 . b_1 b_2 \dots b_n$ represents the number

$$a_m 10^m + a_{m-1} 10^{m-1} + \cdots + a_0 10^0 + \frac{b_1}{10^1} + \frac{b_2}{10^2} + \cdots + \frac{b_n}{10^n}$$

The integer part or integral part of a decimal numeral is the integer written to the left of the decimal separator. For a non-negative decimal numeral, it is the largest integer that is not greater than the decimal. The part from the decimal separator to the right is the fractional part, which equals the difference between the numeral and its integer part.

When the integral part of a numeral is zero, it may occur, typically in computing, that the integer part is not written (for example, .1234, instead of 0.1234). In normal writing, this is generally avoided, because of the risk of confusion between the decimal mark and other punctuation.

In brief, the contribution of each digit to the value of a number depends on its position in the numeral. That is, the decimal system is a positional numeral system.

Decimal fractions

Decimal fractions (sometimes called decimal numbers, especially in contexts involving explicit fractions) are the rational numbers that may be expressed as a fraction whose denominator is a power of ten. For example, the decimals 0.8 , 14.89 , 0.00024 , 1.618 , 3.14159 represent the fractions 8/10, 1489/100, 24/100000, +1618/1000 and +314159/100000, and are therefore decimal numbers.

More generally, a decimal with n digits after the separator (a point or comma) represents the fraction with denominator 10^n, whose numerator is the integer obtained by removing the separator.

It follows that a number is a decimal fraction if and only if it has a finite decimal representation.

Expressed as a fully reduced fraction, the decimal numbers are those whose denominator is a product of a power of 2 and a power of 5. Thus the smallest denominators of decimal numbers are

$$1 = 2^0 \cdot 5^0, 2 = 2^1 \cdot 5^0, 4 = 2^2 \cdot 5^0, 5 = 2^0 \cdot 5^1, 8 = 2^3 \cdot 5^0, 10 = 2^1 \cdot 5^1, 16 = 2^4 \cdot 5^0, 25 = 2^0 \cdot 5^2, \ldots$$

DECIMAL PLACE VALUE

A decimal number consists of a whole number and a fractional part, separated by a decimal point.

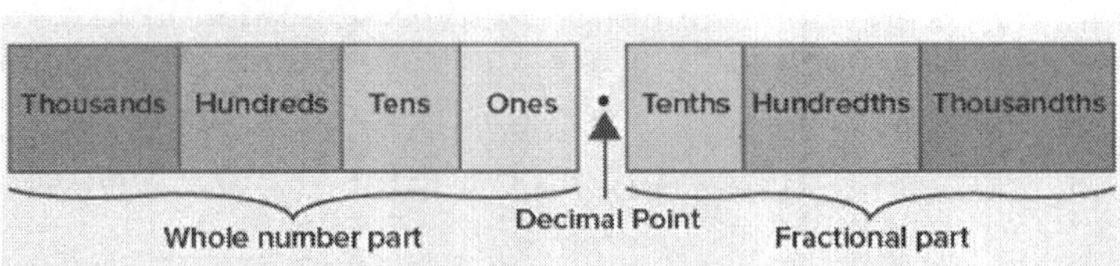

Place Value of A Digit

Depending upon the position of a digit in a number, it has a value called its place value. For example, the place value of the digit 6 in the number 1673 is 600 as 6 is in the hundreds place. However, if we interchange the digits 6 and 7 in the number 1673, we get a new number 1763. In 1763 the place value of the digit 6 is 60 as it is in the tens place.

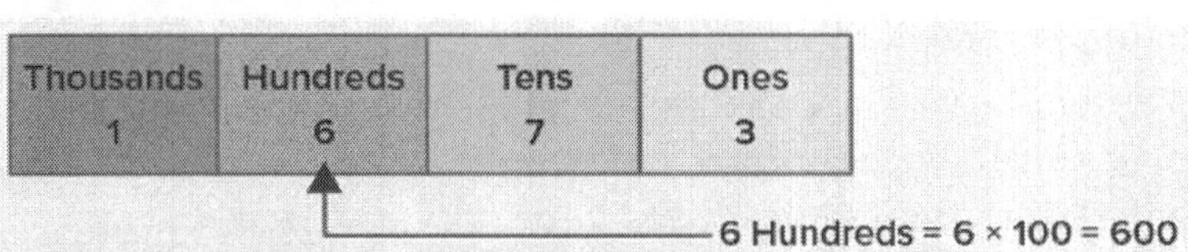

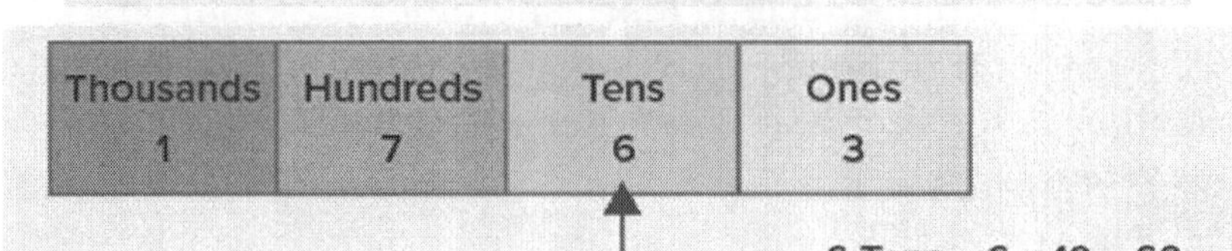

We can use a decimal place value chart to find the place values of the digits in a decimal number.

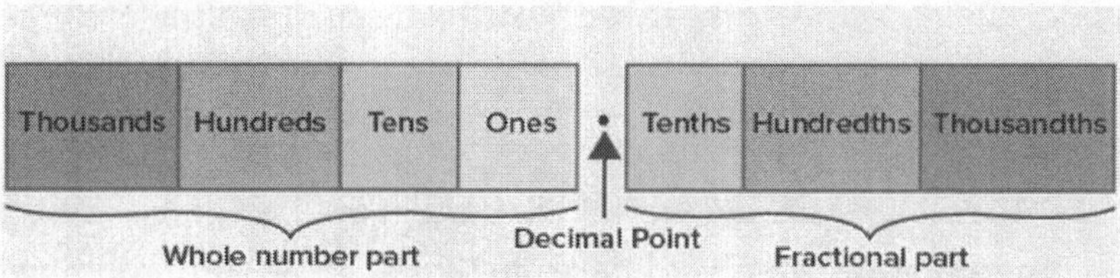

A decimal place value chart helps to find the place value of the digits in a decimal number.

Example: Write the place value of the digits 2 and 4 in the number 326.471

First, write the number in a decimal place value chart.

Then, look at the place of the digit and find its place value.

The digit 2 is in the tens place. Therefore, its place value is 2 tens or 20.

The digit 4 is in the tenths place. Therefore, its place value is 4 tenths or 0.4

Example: Find the place value of the underlined digits in the number 4532.079

In the number 4532.079:

4 is at the thousands place. So, its place value is 4 thousands or 4000

0 is at the tenths place. So, its place value is 0 tenths or 0

9 is at the thousandths place. So, its place value is 9 thousandths or 0.009

Real number approximation

Decimal numerals do not allow an exact representation for all real numbers, e.g. for the real number π. Nevertheless, they allow approximating every real number with any desired accuracy, e.g., the decimal 3.14159 approximates the real π, being less than $10^{''5}$ off; so decimals are widely used in science, engineering and everyday life.

More precisely, for every real number x and every positive integer n, there are two decimals L and u with at most n digits after the decimal mark such that L d" x d" u and (u " L) = $10^{''n}$.

Numbers are very often obtained as the result of measurement. As measurements are subject to measurement uncertainty with a known upper bound, the result of a measurement is well-represented by a decimal with n digits after the decimal mark, as soon as the absolute measurement error is bounded from above by $10^{''n}$. In practice, measurement results are often given with a certain number of digits after the decimal point, which indicate the error bounds. For example, although 0.080 and 0.08 denote the same number, the decimal numeral 0.080 suggests a measurement with an error less than 0.001, while the numeral 0.08 indicates an absolute error bounded by 0.01. In both cases, the true value of the measured quantity could be, for example, 0.0803 or 0.0796.

DECIMAL COMPUTATION

Most modern computer hardware and software systems commonly use a binary representation internally (although many early computers, such as the ENIAC or the IBM 650, used decimal representation internally). For external use by computer specialists, this binary representation is sometimes presented in the related octal or hexadecimal systems.

For most purposes, however, binary values are converted to or from the equivalent decimal values for presentation to or input from humans; computer programs express literals in decimal by default. (123.1, for example, is written as such in a computer program, even though many computer languages are unable to encode that number precisely.)

Both computer hardware and software also use internal representations which are effectively decimal for storing decimal values and doing arithmetic. Often this arithmetic is done on data which are encoded using some variant of binary-coded decimal, especially in database implementations, but there are other decimal representations in use (including decimal floating point such as in newer revisions of the IEEE 754 Standard for Floating-Point Arithmetic).

Decimal arithmetic is used in computers so that decimal fractional results of adding (or subtracting) values with a fixed length of their fractional part always are computed to this same length of precision. This is especially important for financial calculations, e.g., requiring in their results integer multiples of the smallest currency unit for book keeping purposes. This is not possible in binary, because the negative powers of 10 have no finite binary fractional representation; and is generally impossible for multiplication (or division). See Arbitrary-precision arithmetic for exact calculations.

History of Decimal

Many ancient cultures calculated with numerals based on ten, sometimes argued due to human hands typically having ten fingers/digits. Standardized weights used in the Indus Valley civilization (c. 3300–1300 BCE) were based on the ratios: 1/20, 1/10, 1/5, 1/2, 1, 2, 5, 10, 20, 50, 100, 200, and 500, while their standardized ruler – the Mohenjo-daro ruler – was divided into ten equal parts. Egyptian hieroglyphs, in evidence since around 3000 BCE, used a purely decimal system, as did the Cretan hieroglyphs (c. 1625Š1500 BCE) of the Minoans whose numerals are closely based on the Egyptian model. The decimal system was handed down to the consecutive Bronze Age cultures of Greece, including Linear A (c. 18th century BCEŠ1450 BCE) and Linear B (c. 1375Š1200 BCE) – the number system of classical Greece also used powers of ten, including, Roman numerals, an intermediate base of 5. Notably, the polymath Archimedes (c. 287–212 BCE) invented a decimal positional system in his Sand Reckoner which was based on 10^8 and later led the German mathematician Carl Friedrich Gauss to lament what heights science would have already reached in his days if Archimedes had fully realized the potential of his ingenious discovery. Hittite hieroglyphs (since 15th century BCE) were also strictly decimal.

Some non-mathematical ancient texts such as the Vedas, dating back to 1700–900 BCE make use of decimals and mathematical decimal fractions.

The Egyptian hieratic numerals, the Greek alphabet numerals, the Hebrew alphabet numerals, the Roman numerals, the Chinese numerals and early Indian Brahmi numerals are all non-positional decimal systems, and required large numbers of symbols. For instance, Egyptian numerals used different symbols for 10, 20 to 90, 100, 200 to 900, 1000, 2000, 3000, 4000, to 10,000. The world's earliest positional decimal system was the Chinese rod calculus.

HISTORY OF DECIMAL FRACTIONS

Decimal fractions were first developed and used by the Chinese in the end of 4th century BCE, and then spread to the Middle East and from there to Europe. The written Chinese decimal fractions were non-positional. However, counting rod fractions were positional.

Qin Jiushao in his book Mathematical Treatise in Nine Sections (1247) denoted 0.96644 by

J. Lennart Berggren notes that positional decimal fractions appear for the first time in a book by the Arab mathematician Abu'l-Hasan al-Uqlidisi written in the 10th century. The Jewish mathematician Immanuel Bonfils used decimal fractions around 1350, anticipating Simon Stevin, but did not develop any notation to represent them. The Persian mathematician Jamshîd al-Kâshî claimed to have discovered decimal fractions himself in the 15th century. Al Khwarizmi introduced fraction to Islamic countries in the early 9th century; a Chinese author has alleged that his fraction presentation was an exact copy of traditional Chinese mathematical fraction from Sunzi Suanjing. This form of fraction with numerator on top and denominator at bottom without a horizontal bar was also used by al-Uqlidisi and by al-Kâshî in his work "Arithmetic Key".

A forerunner of modern European decimal notation was introduced by Simon Stevin in the 16th century.

John Napier introduced using the period (.) to separate the integer part of a decimal number from the fractional part in his book on constructing tables of logarithms, published posthumously in 1620.: p. 8, archive p. 32)

NATURAL LANGUAGES

A method of expressing every possible natural number using a set of ten symbols emerged in India. Several Indian languages show a straightforward decimal system. Many Indo-Aryan and Dravidian languages have numbers between 10 and 20 expressed in a regular pattern of addition to 10.

The Hungarian language also uses a straightforward decimal system. All numbers between 10 and 20 are formed regularly (e.g. 11 is expressed as "tizenegy" literally "one on ten"), as with those between 20 and 100 (23 as "huszonhárom" = "three on twenty").

A straightforward decimal rank system with a word for each order, and in which 11 is expressed as ten-one and 23 as two-ten-three, and 89,345 is expressed as 8 (ten thousands) 9 (thousand) 3 (hundred) 4 (tens) AS 5 is found in Chinese, and in Vietnamese with a few irregularities. Japanese, Korean, and Thai have imported the Chinese decimal system. Many other languages with a decimal system have special words for the numbers between 10 and 20, and decades. For example, in English 11 is "eleven" not "ten-one" or "one-teen".

Incan languages such as Quechua and Aymara have an almost straightforward decimal system, in which 11 is expressed as ten with one and 23 as two-ten with three.

Some psychologists suggest irregularities of the English names of numerals may hinder children's counting ability.

Other bases

Some cultures do, or did, use other bases of numbers.

- Pre-Columbian Mesoamerican cultures such as the Maya used a base-20 system (perhaps based on using all twenty fingers and toes).
- The Yuki language in California and the Pamean languages in Mexico have octal (base-8) systems because the speakers count using the spaces between their fingers rather than the fingers themselves.
- The existence of a non-decimal base in the earliest traces of the Germanic languages is attested by the presence of words and glosses meaning that the count is in decimal (cognates to "ten-count" or "tenty-wise"); such would be expected if normal counting is not decimal, and unusual

if it were. Where this counting system is known, it is based on the "long hundred" = 120, and a "long thousand" of 1200. The descriptions like "long" only appear after the "small hundred" of 100 appeared with the Christians. Gordon's Introduction to Old Norse p. 293, gives number names that belong to this system. An expression cognate to 'one hundred and eighty' translates to 200, and the cognate to 'two hundred' translates to 240. Goodare details the use of the long hundred in Scotland in the Middle Ages, giving examples such as calculations where the carry implies i C (i.e. one hundred) as 120, etc. That the general population were not alarmed to encounter such numbers suggests common enough use. It is also possible to avoid hundred-like numbers by using intermediate units, such as stones and pounds, rather than a long count of pounds. Goodare gives examples of numbers like vii score, where one avoids the hundred by using extended scores. There is also a paper by W.H. Stevenson, on 'Long Hundred and its uses in England'.

- Many or all of the Chumashan languages originally used a base-4 counting system, in which the names for numbers were structured according to multiples of 4 and 16.
- Many languages use quinary (base-5) number systems, including Gumatj, Nunggubuyu, Kuurn Kopan Noot and Saraveca. Of these, Gumatj is the only true 5–25 language known, in which 25 is the higher group of 5.
- Some Nigerians use duodecimal systems. So did some small communities in India and Nepal, as indicated by their languages.
- The Huli language of Papua New Guinea is reported to have base-15 numbers. Ngui means 15, ngui ki means 15 × 2 = 30, and ngui ngui means 15 × 15 = 225.
- Umbu-Ungu, also known as Kakoli, is reported to have base-24 numbers. Tokapu means 24, tokapu talu means 24 × 2 = 48, and tokapu tokapu means 24 × 24 = 576.
- Ngiti is reported to have a base-32 number system with base-4 cycles.
- The Ndom language of Papua New Guinea is reported to have base-6 numerals. Mer means 6, mer an thef means 6 × 2 = 12, nif means 36, and nif thef means 36×2 = 72.

Significant and Insignifican

Significant figures (also known as the significant digits, precision or resolution) of a number in positional notation are digits in the number that are reliable and necessary to indicate the quantity of something.

If a number expressing the result of a measurement (e.g., length, pressure, volume, or mass) has more digits than the number of digits allowed by the measurement resolution, then only as many digits as allowed by the measurement resolution are reliable, and so only these can be significant figures.

For example, if a length measurement gives 114.8 mm while the smallest interval between marks on the ruler used in the measurement is 1 mm, then the first three digits (1, 1, and 4, showing 114 mm) are certain and so they are significant figures. Digits which are uncertain but reliable are also considered significant figures. In this example, the last digit (8, which adds 0.8 mm) is also considered a significant figure even though there is uncertainty in it.

Another example is a volume measurement of 2.98 L with an uncertainty of ± 0.05 L. The actual volume is somewhere between 2.93 L and 3.03 L. Even when some of the digits are not certain, as long as they are reliable, they are considered significant because they indicate the actual volume within the acceptable degree of uncertainty. In this example the actual volume might be 2.94 L or might instead be 3.02 L. And so all three are significant figures.

The following digits are not significant figures.

- All leading zeros. For example, 013 kg has two significant figures, 1 and 3, and the leading zero is not significant since it is not necessary to indicate the mass; 013 kg = 13 kg so 0 is not necessary. In the case of 0.056 m there are two insignificant leading zeros since 0.056 m = 56 mm and so the leading zeros are not necessary to indicate the length.
- Trailing zeros when they are merely placeholders. For example, the trailing zeros in 1500 m as a length measurement are not significant if they are just placeholders for ones and tens places as the measurement resolution is 100 m. In this case, 1500 m means the length to measure is close to 1500 m rather than saying that the length is exactly 1500 m.
- Spurious digits, introduced by calculations resulting in a number with a greater precision than the precision of the used data in the calculations, or in a measurement reported to a greater precision than the measurement resolution.

Of the significant figures in a number, the most significant is the digit with the highest exponent value (simply the left-most significant figure), and the least significant is the digit with the lowest exponent value (simply the right-most significant figure). For example, in the number "123", the "1" is the most significant figure as it counts hundreds (10^2), and "3" is the least significant figure as it counts ones (10^0).

Significance arithmetic is a set of approximate rules for roughly maintaining significance throughout a computation. The more sophisticated scientific rules are known as propagation of uncertainty.

Numbers are often rounded to avoid reporting insignificant figures. For example, it would create false precision to express a measurement as 12.34525 kg if the scale was only measured to the nearest gram. In this case, the significant figures are the first 5 digits from the left-most digit (1, 2, 3, 4, and 5), and the number needs to be rounded to the significant figures so that it will be 12.345 kg as the reliable value. Numbers can also be rounded merely for simplicity rather than to indicate a precision of measurement, for example, in order to make the numbers faster to pronounce in news broadcasts.

Radix 10 (base-10, decimal numbers) is assumed in the following.

Identifying significant figures

Rules to identify significant figures in a number

Note that identifying the significant figures in a number requires knowing which digits are reliable (e.g., by knowing the measurement or reporting resolution with which the number is obtained or processed) since only reliable digits can be significant; e.g., 3 and 4 in 0.00234 g are not significant if the measurable smallest weight is 0.001 g.

- Non-zero digits within the given measurement or reporting resolution are significant.
  - o 91 has two significant figures (9 and 1) if they are measurement-allowed digits.
  - o 123.45 has five significant digits (1, 2, 3, 4 and 5) if they are within the measurement resolution. If the resolution is 0.1, then the last digit 5 is not significant.
- Zeros between two significant non-zero digits are significant (significant trapped zeros).
  - o 101.12003 consists of eight significant figures if the resolution is to 0.00001.
  - o 125.340006 has seven significant figures if the resolution is to 0.0001: 1, 2, 5, 3, 4, 0, and 0.
- Zeros to the left of the first non-zero digit (leading zeros) are not significant.
  - o If a length measurement gives 0.052 km, then 0.052 km = 52 m so 5 and 2 are only significant; the leading zeros appear or disappear, depending on which unit is used, so they are not necessary to indicate the measurement scale.
  - o 0.00034 has 4 significant zeros if the resolution is 0.001. (3 and 4 are beyond the resolution so are not significant.)

- Zeros to the right of the last non-zero digit (trailing zeros) in a number with the decimal point are significant if they are within the measurement or reporting resolution.
  - o 1.200 has four significant figures (1, 2, 0, and 0) if they are allowed by the measurement resolution.
  - o 0.0980 has three significant digits (9, 8, and the last zero) if they are within the measurement resolution.
  - o 120.000 consists of six significant figures (1, 2, and the four subsequent zeroes) except for the last zero If the resolution is to 0.01.
- Trailing zeros in an integer may or may not be significant, depending on the measurement or reporting resolution.
  - o 45,600 has 3, 4 or 5 significant figures depending on how the last zeros are used. For example, if the length of a road is reported as 45600 m without information about the reporting or measurement resolution, then it is not clear if the road length is precisely measured as 45600 m or if it is a rough estimate. If it is the rough estimation, then only the first three non-zero digits are significant since the trailing zeros are neither reliable nor necessary; 45600 m can be expressed as 45.6 km or as 4.56×10^4 m in scientific notation, and neither expression requires the trailing zeros.
- An exact number has an infinite number of significant figures.
  - o If the number of apples in a bag is 4 (exact number), then this number is 4.0000... (with infinite trailing zeros to the right of the decimal point). As a result, 4 does not impact the number of significant figures or digits in the result of calculations with it.
- A mathematical or physical constant has significant figures to its known digits.
  - o π, as the ratio of the circumference to the diameter of a circle, is 3.14159265358979323... known to 50 trillion digits calculated as of 2020-01-29, and that calculated 'π' approximation has that many significant digits, while in practical applications far fewer are used (and π itself has infinite significant digits, as all irrational numbers do). Often 3.14 is used in numerical calculations, i.e. 3 significant decimal digits, with 7 correct binary digits (while the more accurate 22/7 is also used, even though it also only amounts to the same 3 significant correct decimal digits, it has 10 correct binary digits), which is a good enough approximation for many practical uses. Most calculators, and computer programs, can handle 3.141592653589793, 16 decimal digits, that is commonly used in computers and used by NASA for "JPL's highest accuracy

calculations, which are for interplanetary navigation". For "the largest size there is: the visible universe [..] you would need 39 or 40 decimal places."

- o The Planck constant is $h = 6.62607015 \times 10^{-34}\, J{\cdot}s$ and is defined as an exact value so that it is more properly defined as $h = 6.62607015(0) \times 10^{-34}\, J{\cdot}s$.

Ways to denote significant figures in an integer with trailing zeros

The significance of trailing zeros in a number not containing a decimal point can be ambiguous. For example, it may not always be clear if the number 1300 is precise to the nearest unit (just happens coincidentally to be an exact multiple of a hundred) or if it is only shown to the nearest hundreds due to rounding or uncertainty. Many conventions exist to address this issue. However, these are not universally used and would only be effective if the reader is familiar with the convention:

- An overline, sometimes also called an overbar, or less accurately, a vinculum, may be placed over the last significant figure; any trailing zeros following this are insignificant. For example, 1300 has three significant figures (and hence indicates that the number is precise to the nearest ten).
- Less often, using a closely related convention, the last significant figure of a number may be underlined; for example, "1300" has two significant figures.
- A decimal point may be placed after the number; for example "1300." indicates specifically that trailing zeros are meant to be significant.

As the conventions above are not in general use, the following more widely recognized options are available for indicating the significance of number with trailing zeros:

- Eliminate ambiguous or non-significant zeros by changing the unit prefix in a number with a unit of measurement. For example, the precision of measurement specified as 1300 g is ambiguous, while if stated as 1.30 kg it is not. Likewise 0.0123 L can be rewritten as 12.3 mL
- Eliminate ambiguous or non-significant zeros by using Scientific Notation: For example, 1300 with three significant figures becomes 1.30×10^3. Likewise 0.0123 can be rewritten as 1.23×10^{Š2}. The part of the representation that contains the significant figures (1.30 or 1.23) is known as the significand or mantissa. The digits in the base and exponent (10^3 or $10^{Š2}$) are considered exact numbers so for these digits, significant figures are irrelevant.
- Explicitly state the number of significant figures (the abbreviation s.f. is sometimes used): For example "20 000 to 2 s.f." or "20 000 (2 sf)".

- State the expected variability (precision) explicitly with a plus–minus sign, as in 20 000 ± 1%. This also allows specifying a range of precision in-between powers of ten.

ROUNDING TO SIGNIFICANT FIGURES

Rounding to significant figures is a more general-purpose technique than rounding to n digits, since it handles numbers of different scales in a uniform way. For example, the population of a city might only be known to the nearest thousand and be stated as 52,000, while the population of a country might only be known to the nearest million and be stated as 52,000,000. The former might be in error by hundreds, and the latter might be in error by hundreds of thousands, but both have two significant figures (5 and 2). This reflects the fact that the significance of the error is the same in both cases, relative to the size of the quantity being measured.

To round a number to n significant figures:

1. If the n + 1 digit is greater than 5 or is 5 followed by other non-zero digits, add 1 to the n digit. For example, if we want to round 1.2459 to 3 significant figures, then this step results in 1.25.
2. If the n + 1 digit is 5 not followed by other digits or followed by only zeros, then rounding requires a tie-breaking rule. For example, to round 1.25 to 2 significant figures:
   - o Round half away from zero (also known as "5/4") rounds up to 1.3. This is the default rounding method implied in many disciplines if the required rounding method is not specified.
   - o Round half to even, which rounds to the nearest even number. With this method, 1.25 is rounded down to 1.2. If this method applies to 1.35, then it is rounded up to 1.4. This is the method preferred by many scientific disciplines, because, for example, it avoids skewing the average value of a long list of values upwards.
3. For an integer in rounding, replace the digits after the n digit with zeros. For example, if 1254 is rounded to 2 significant figures, then 5 and 4 are replaced to 0 so that it will be 1300. For a number with the decimal point in rounding, remove the digits after the n digit. For example, if 14.895 is rounded to 3 significant figures, then the digits after 8 are removed so that it will be 14.9.

In financial calculations, a number is often rounded to a given number of places. For example, to two places after the decimal separator for many world currencies. This is done because greater precision is immaterial, and usually it is not possible to settle a debt of less than the smallest currency unit.

In UK personal tax returns, income is rounded down to the nearest pound, whilst tax paid is calculated to the nearest penny.

As an illustration, the decimal quantity 12.345 can be expressed with various numbers of significant figures or decimal places. If insufficient precision is available then the number is rounded in some manner to fit the available precision. The following table shows the results for various total precision at two rounding ways (N/A stands for Not Applicable).

WRITING UNCERTAINTY AND IMPLIED UNCERTAINTY

Significant figures in writing uncertainty

It is recommended for a measurement result to include the measurement uncertainty such as $x_{best} \pm \sigma_x$ where x_{best} and ó$_x$ are the best estimate and uncertainty in the measurement respectively. x_{best} can be the average of measured values and ó$_x$ can be the standard deviation or a multiple of the measurement deviation. The rules to write $x_{best} \pm \sigma_x$ are:

- ó$_x$ has only one or two significant figures as more precise uncertainty has no meaning.
  - o 1.79 ± 0.06 (correct), 1.79 ± 0.96 (correct), 1.79 ± 1.96 (incorrect).
- The digit positions of the last significant figures in x_{best} and ó$_x$ are the same, otherwise the consistency is lost. For example, in 1.79 ± 0.067 (incorrect), it does not make sense to have more accurate uncertainty than the best estimate. 1.79 ± 0.9 (incorrect) also does not make sense since the rounding guideline for addition and subtraction below tells that the edges of the true value range are 2.7 and 0.9, that are less accurate than the best estimate.
  - o 1.79 ± 0.06 (correct), 1.79 ± 0.96 (correct), 1.79 ± 0.067 (incorrect), 1.79 ± 0.9 (incorrect).

Implied uncertainty

In chemistry (and may also be for other scientific branches), uncertainty may be implied by the last significant figure if it is not explicitly expressed. The implied uncertainty is ± the half of the minimum scale at the last significant figure position. For example, if the volume of water in a bottle is reported as 3.78 L without mentioning uncertainty, then ± 0.005 L measurement uncertainty may be implied. If 2.97 ± 0.07 kg, so the actual weight is somewhere in 2.90 to 3.04 kg, is measured and it is desired to report it with a single number, then 3.0 kg is the best number to report since its implied uncertainty ± 0.05 kg tells the weight range of 2.95 to 3.05 kg that is close to the measurement range. If 2.97 ± 0.09 kg, then 3.0 kg is still the best since, if 3 kg is reported then its implied uncertainty ± 0.5 tells the range of 2.5 to 3.5 kg that is too wide in comparison with the measurement range.

If there is a need to write the implied uncertainty of a number, then it can be written as x ± ó$_x$ with stating it as the implied uncertainty (to prevent readers from recognizing it as the measurement uncertainty), where x and ó$_x$ are the number with an extra zero digit (to follow the rules to write uncertainty above) and the implied uncertainty of it respectively. For example, 6 kg with the implied uncertainty ± 0.5 kg can be stated as 6.0 ± 0.5 kg.

ARITHMETIC

As there are rules to determine the significant figures in directly measured quantities, there are also guidelines (not rules) to determine the significant figures in quantities calculated from these measured quantities.

Significant figures in measured quantities are most important in the determination of significant figures in calculated quantities with them. A mathematical or physical constant (e.g., π in the formula for the area of a circle with radius r as πr^2) has no effect on the determination of the significant figures in the result of a calculation with it if its known digits are equal to or more than the significant figures in the measured quantities used in the calculation. An exact number such as ½ in the formula for the kinetic energy of a mass m with velocity v as ½mv^2 has no bearing on the significant figures in the calculated kinetic energy since its number of significant figures is infinite (0.500000...).

The guidelines described below are intended to avoid a calculation result more precise than the measured quantities, but it does not ensure the resulted implied uncertainty close enough to the measured uncertainties. This problem can be seen in unit conversion. If the guidelines give the implied uncertainty too far from the measured ones, then it may be needed to decide significant digits that give comparable uncertainty.

Multiplication and division

For quantities created from measured quantities via multiplication and division, the calculated result should have as many significant figures as the least number of significant figures among the measured quantities used in the calculation. For example,

- 1.234 × 2 = 2.468 HŠ 2
- 1.234 × 2.0 = 2.468 HŠ 2.5
- 0.01234 × 2 = 0.02468 HŠ 0.02

with one, two, and one significant figures respectively. (2 here is assumed not an exact number.) For the first example, the first multiplication factor has four significant figures and the second has one significant figure. The factor with the fewest or least significant figures is the second one with only one, so the final calculated result should also have one significant figure.

Exception

For unit conversion, the implied uncertainty of the result can be unsatisfactorily higher than that in the previous unit if this rounding guideline is followed; For example, 8 inch has the implied uncertainty of ± 0.5 inch = ± 1.27 cm. If it is converted to the centimetre scale and the rounding guideline for multiplication and division is followed, then 20.32 cm HŠ 20 cm with the implied uncertainty of ± 5 cm. If this implied uncertainty is considered as too underestimated, then more proper significant digits in the unit conversion result may be 20.32 cm HŠ 20. cm with the implied uncertainty of ± 0.5 cm.

Another exception of applying the above rounding guideline is to multiply a number by an integer, such as 1.234 × 9. If the above guideline is followed, then the result is rounded as 1.234 × 9.000.... = 11.106 HŠ 11.11. However, this multiplication is essentially adding 1.234 to itself 9 times such as 1.234 + 1.234 + ... + 1.234 so the rounding guideline for addition and subtraction described below is more proper rounding approach. As a result, the final answer is 1.234 + 1.234 + ... + 1.234 = 11.106 = 11.106 (one significant digit increase).

Addition and subtraction

For quantities created from measured quantities via addition and subtraction, the last significant figure position (e.g., hundreds, tens, ones, tenths, hundredths, and so forth) in the calculated result should be the same as the leftmost or largest digit position among the last significant figures of the measured quantities in the calculation. For example,

- 1.234 + 2 = 3.234 H" 3
- 1.234 + 2.0 = 3.234 H" 3.2
- 0.01234 + 2 = 2.01234 H" 2

with the last significant figures in the ones place, tenths place, and ones place respectively. (2 here is assumed not an exact number.) For the first example, the first term has its last significant figure in the thousandths place and the second term has its last significant figure in the ones place. The leftmost or largest digit position among the last significant figures of these terms is the ones place, so the calculated result should also have its last significant figure in the ones place.

The rule to calculate significant figures for multiplication and division are not the same as the rule for addition and subtraction. For multiplication and division, only the total number of significant figures in each of the factors in the calculation matters; the digit position of the last significant figure in each factor is irrelevant. For addition and subtraction, only the digit position of the last significant figure in each of the terms in the calculation matters; the total number of significant figures in each term is irrelevant. However, greater accuracy will often be obtained if some non-significant digits are maintained in intermediate results which are used in subsequent calculations.

Logarithm and antilogarithm

The base-10 logarithm of a normalized number (i.e., a × 10^b with 1 d" a < 10 and b as an integer), is rounded such that its decimal part (called mantissa) has as many significant figures as the significant figures in the normalized number.

- $\log_{10}(3.000 \times 10^4) = \log_{10}(10^4) + \log_{10}(3.000)$ = 4.000000... (exact number so infinite significant digits) + 0.4771212547... = 4.4771212547 H" 4.4771.

When taking the antilogarithm of a normalized number, the result is rounded to have as many significant figures as the significant figures in the decimal part of the number to be antiloged.

- $10^{4.4771} = 29998.5318119... = 30000 = 3.000 \times 10^4$.

Round only on the final calculation result

When performing multiple stage calculations, do not round intermediate stage calculation results; keep as many digits as is practical (at least one more digit than the rounding rule allows per stage) until the end of all the calculations to avoid cumulative rounding errors while tracking or recording the significant figures in each intermediate result. Then, round the final result, for example, to the fewest number of significant figures (for multiplication or division) or leftmost last significant digit position (for addition or subtraction) among the inputs in the final calculation.

- (2.3494 + 1.345) × 1.2 = 3.6944 × 1.2 = 4.43328 HŠ 4.4.
- (2.3494 × 1.345) + 1.2 = 3.159943 + 1.2 = 4.359943 HŠ 4.4.

Estimating an extra digit

When using a ruler, initially use the smallest mark as the first estimated digit. For example, if a ruler's smallest mark is 0.1 cm, and 4.5 cm is read, then it is 4.5 (±0.1 cm) or 4.4 cm to 4.6 cm as to the smallest mark interval. However, in practice a measurement can usually be estimated by eye to closer than the interval between the ruler's smallest mark, e.g. in the above case it might be estimated as between 4.51 cm and 4.53 cm.

It is also possible that the overall length of a ruler may not be accurate to the degree of the smallest mark, and the marks may be imperfectly spaced within each unit. However assuming a normal good quality ruler, it should be possible to estimate tenths between the nearest two marks to achieve an extra decimal place of accuracy. Failing to do this adds the error in reading the ruler to any error in the calibration of the ruler.

ESTIMATION IN STATISTIC

When estimating the proportion of individuals carrying some particular characteristic in a population, from a random sample of that population, the number of significant figures should not exceed the maximum precision allowed by that sample size.

Relationship to accuracy and precision in measurement

Traditionally, in various technical fields, "accuracy" refers to the closeness of a given measurement to its true value; "precision" refers to the stability of that measurement when repeated many times. Hoping to reflect the way in which the term "accuracy" is actually used in the scientific community, there is a recent standard, ISO 5725, which keeps the same definition of precision but defines the term "trueness" as the closeness of a given measurement to its true value and uses the term "accuracy" as the combination of trueness and precision. (See the accuracy and precision article for a full discussion.) In either case, the number of significant figures roughly corresponds to precision, not to accuracy or the newer concept of trueness.

Comparing Decimals

Comparing decimals means finding out the larger and the smaller decimal number in a given set of numbers. Decimal numbers can be compared in the same way as we compare other numbers. However, we need to remember that the digits that are placed after the decimal point also need to be considered. These digits have place values starting from tenths, followed by hundredths, then thousandths, and so on. Let us learn more about comparing decimals in this article.

Comparing decimals is similar to comparing other whole numbers in which we start comparing the digits with the greatest place value. We place the given decimal numbers in a place value chart and start the comparison. If the digits on the greatest place value are the same, we move on to the digits in the next place to the right. We keep comparing digits until we reach the digits that are different. Let us understand this with the help of the following example.

Example: Compare 0.64 and 0.362

Solution:

- Step 1: Compare the whole number part, which is the ones digit. If the numbers are the same, go to the next step. In this case, the ones digits has 0 in both the numbers. So, we move on to the next place to the right.
- Step 2: Compare the tenths place, which is the place to the right of the decimal point. When we compare the value in the tenths place, we see that 6 is greater than 3. At this step itself, we get to know that 0.64 is greater than 0.362. Therefore, we do not need to move on to the hundredths digit for further comparison.
- Step 3: Hence, we conclude that 0.64 > 0.362

Comparing Decimals and Fractions

When we need to compare decimals and fractions, we first convert the given fraction to a decimal number and then compare the numbers using the same procedure.

Example: Compare 3/4 and 0.728

Solution: First, let us convert 3/4 to a decimal number by dividing 3 by 4. So, 3 ÷ 4 = 0.75. Now, we have both the numbers in the decimal form. So, let us compare 0.75 and 0.728 using the method given above.

- Step 1: Compare the whole number part, which is the ones digit. In this case, the ones digits has 0 in both the numbers. So, we will move on to the next place value to the right.
- Step 2: Compare the tenths place, which is the first place to the right of the decimal point. When we compare the value on the tenths digit, we see that both the numbers have 7. So we move on to the hundredths place.
- Step 3: Compare the hundredths place. Now, when we compare the value at the hundredths place, we see that 5 is greater than 2. At this step, we get to know that 0.75 is greater than 0.728. Therefore, we do not need to move on to the thousandths digit for further comparison.
- Step 3: Hence, we conclude that 0.75 > 0.728 which means 3/4 > 0.728

Comparing Decimals to Hundredths place

We have already seen that we compare decimal numbers starting from the whole number part and then we move on to the digits given after the decimal point. Let us see how to compare decimals to the hundredths place.

Example: Compare 8.362 and 8.391 to find the greater number.

Solution: Let us follow the same steps as shown above.

- Step 1: Compare the whole number part. In this case, the whole number part has 8 in both the numbers. So, we will move on to the next place value.

- Step 2: Compare the tenths place, which is the first place to the right of the decimal point. When we compare the value on the tenths digit, we see that both the numbers have 3. At this point, we move on to the hundredths place.
- Step 3: Compare the hundredths place. Now, when we compare the value at the hundredths place, we see that 9 is greater than 6. So, we get to know that 8.391 is greater than 8.362. Therefore, we do not need to move on to the thousandths digit for further comparison.
- Step 3: Hence, we conclude that 8.362 < 8.391

Percent

The term 'per cent' means 'out of a hundred'. In mathematics, percentages are used like fractions and decimals, as ways to describe parts of a whole. When you are using percentages, the whole is considered to be made up of a hundred equal parts. The symbol % is used to show that a number is a percentage, and less commonly the abbreviation 'pct' may be used.

Examples

For example, 45% (read as "forty-five percent") is equal to the fraction 45/100, the ratio 45:55 (or 45:100 when comparing to the total rather than the other portion), or 0.45. Percentages are often used to express a proportionate part of a total.

(Similarly, one can also express a number as a fraction of 1000, using the term "per mille" or the symbol "‰".)

Example 1

If 50% of the total number of students in the class are male, that means that 50 out of every 100 students are male. If there are 500 students, then 250 of them are male.

Example 2

An increase of $0.15 on a price of $2.50 is an increase by a fraction of 0.15/2.50 = 0.06. Expressed as a percentage, this is a 6% increase.

While many percentage values are between 0 and 100, there is no mathematical restriction and percentages may take on other values. For example, it is common to refer to 111% or Š35%, especially for percent changes and comparisons.

History

In Ancient Rome, long before the existence of the decimal system, computations were often made in fractions in the multiples of 1/100. For example, Augustus levied a tax of 1/100 on goods sold at auction known as centesima rerum venalium. Computation with these fractions was equivalent to computing percentages.

As denominations of money grew in the Middle Ages, computations with a denominator of 100 became increasingly standard, such that from the late 15th century to the early 16th century, it became common for arithmetic texts to include such computations. Many of these texts applied these methods to profit and loss, interest rates, and the Rule of Three. By the 17th century, it was standard to quote interest rates in hundredths.

CALCULATIONS

The percent value is computed by multiplying the numeric value of the ratio by 100. For example, to find 50 apples as a percentage of 1250 apples, one first computes the ratio 50/1250 = 0.04, and then multiplies by 100 to obtain 4%. The percent value can also be found by multiplying first instead of later, so in this example, the 50 would be multiplied by 100 to give 5,000, and this result would be divided by 1250 to give 4%.

To calculate a percentage of a percentage, convert both percentages to fractions of 100, or to decimals, and multiply them. For example, 50% of 40% is:

$50/100 \times 40/100 = 0.50 \times 0.40 = 0.20 = 20/100 = 20\%$.

It is not correct to divide by 100 and use the percent sign at the same time; it would literally imply division by 10,000. For example, 25% = 25/100 = 0.25, not 25%/100, which actually is 25D 100/100 = 0.0025. A term such as 100/100% would also be incorrect, since it would be read as 1 percent, even if the intent was to say 100%.

Whenever communicating about a percentage, it is important to specify what it is relative to (i.e., what is the total that corresponds to 100%). The following problem illustrates this point.

In a certain college 60% of all students are female, and 10% of all students are computer science majors. If 5% of female students are computer science majors, what percentage of computer science majors are female?

We are asked to compute the ratio of female computer science majors to all computer science majors. We know that 60% of all students are female, and among these 5% are computer science majors, so we conclude that $60/100 \times 5/100 = 3/100$ or 3% of all students are female computer science majors. Dividing this by the 10% of all students that are computer science majors, we arrive at the answer: 3%/10% = 30/100 or 30% of all computer science majors are female.

This example is closely related to the concept of conditional probability.

Percentage increase and decrease

Due to inconsistent usage, it is not always clear from the context what a percentage is relative to. When speaking of a "10% rise" or a "10% fall" in a quantity, the usual interpretation is that this is relative to the initial value of that quantity. For example, if an item is initially priced at $200 and the price rises 10% (an increase of $20), the new price will be $220. Note that this final price is 110% of the initial price (100% + 10% = 110%).

Some other examples of percent changes:

- An increase of 100% in a quantity means that the final amount is 200% of the initial amount (100% of initial + 100% of increase = 200% of initial). In other words, the quantity has doubled.
- An increase of 800% means the final amount is 9 times the original (100% + 800% = 900% = 9 times as large).
- A decrease of 60% means the final amount is 40% of the original (100% – 60% = 40%).
- A decrease of 100% means the final amount is zero (100% – 100% = 0%).

In general, a change of x percent in a quantity results in a final amount that is 100 + x percent of the original amount (equivalently, (1 + 0.01x) times the original amount).

COMPOUNDING PERCENTAGES

Percent changes applied sequentially do not add up in the usual way. For example, if the 10% increase in price considered earlier (on the $200 item, raising its price to $220) is followed by a 10% decrease in the price (a decrease of $22), then the final price will be $198—not the original price of $200. The reason for this apparent discrepancy is that the two percent changes (+10% and ″10%) are measured relative to different quantities ($200 and $220, respectively), and thus do not "cancel out".

In general, if an increase of x percent is followed by a decrease of x percent, and the initial amount was p, the final amount is p(1 + 0.01x)(1 ″ 0.01x) = p(1 ″ (0.01x)^2); hence the net change is an overall decrease by x percent of x percent (the square of the original percent change when expressed as a decimal number). Thus, in the above example, after an increase and decrease of x = 10 percent, the final amount, $198, was 10% of 10%, or 1%, less than the initial amount of $200. The net change is the same for a decrease of x percent, followed by an increase of x percent; the final amount is p(1 - 0.01x)(1 + 0.01x) = p(1 ″ (0.01x)^2).

This can be expanded for a case where one does not have the same percent change. If the initial amount p leads to a percent change x, and the second percent change is y, then the final amount is p(1 + 0.01x)(1 + 0.01y). To change the above example, after an increase of x = 10 percent and decrease of y = "5 percent, the final amount, $209, is 4.5% more than the initial amount of $200.

As shown above, percent changes can be applied in any order and have the same effect.

In the case of interest rates, a very common but ambiguous way to say that an interest rate rose from 10% per annum to 15% per annum, for example, is to say that the interest rate increased by 5%, which could theoretically mean that it increased from 10% per annum to 10.05% per annum. It is clearer to say that the interest rate increased by 5 percentage points (pp). The same confusion between the different concepts of percent(age) and percentage points can potentially cause a major misunderstanding when journalists report about election results, for example, expressing both new results and differences with earlier results as percentages. For example, if a party obtains 41% of the vote and this is said to be a 2.5% increase, does that mean the earlier result was 40% (since 41 = 40 × (1 + 2.5/100)) or 38.5% (since 41 = 38.5 + 2.5)?

In financial markets, it is common to refer to an increase of one percentage point (e.g. from 3% per annum to 4% per annum) as an increase of "100 basis points".

Word and symbol

In British English, percent is usually written as two words (per cent), although percentage and percentile are written as one word. In American English, percent is the most common variant (but per mille is written as two words).

In the early 20th century, there was a dotted abbreviation form "per cent.", as opposed to "per cent". The form "per cent." is still in use in the highly formal language found in certain documents like commercial loan agreements (particularly those subject to, or inspired by, common law), as well as in the Hansard transcripts of British Parliamentary proceedings. The term has been attributed to Latin per centum. The concept of considering values as parts of a hundred is originally Greek. The symbol for percent (%) evolved from a symbol abbreviating the Italian per cento. In some other languages, the form procent or prosent is used instead. Some languages use both a word derived from percent and an expression in that language meaning the same thing, e.g. Romanian procent and la sutã (thus, 10% can be read or sometimes written ten for [each] hundred, similarly with the English one out of ten). Other abbreviations are rarer, but sometimes seen.

Grammar and style guides often differ as to how percentages are to be written. For instance, it is commonly suggested that the word percent (or per cent) be spelled out in all texts, as in "1 percent" and not "1%". Other guides prefer the word to be written out in humanistic texts, but the symbol to be used in scientific texts. Most guides agree that they always be written with a numeral, as in "5 percent" and not "five percent", the only exception being at the beginning of a sentence: "Ten percent of all writers love style guides." Decimals are also to be used instead of fractions, as in "3.5 percent of the gain" and not "3+1D 2 percent of the gain". However the titles of bonds issued by governments and other issuers use the fractional form, e.g. "3+1D 2% Unsecured Loan Stock 2032 Series 2". (When interest rates are very low, the number 0 is included if the interest rate is less than 1%, e.g. "0+3D 4% Treasury Stock", not "3D 4% Treasury Stock".) It is also widely accepted to use the percent symbol (%) in tabular and graphic material.

In line with common English practice, style guides—such as The Chicago Manual of Style—generally state that the number and percent sign are written without any space in between. However, the International System of Units and the ISO 31-0 standard require a space.

Other uses

The word "percentage" is often a misnomer in the context of sports statistics, when the referenced number is expressed as a decimal proportion, not a percentage: "The Phoenix Suns' Shaquille O'Neal led the NBA with a .609 field goal percentage (FG%) during the 2008–09 season." (O'Neal made 60.9% of his shots, not 0.609%.) Likewise, the winning percentage of a team, the fraction of matches that the club has won, is also usually expressed as a decimal proportion; a team that has a .500 winning percentage has won 50% of their matches. The practice is probably related to the similar way that batting averages are quoted.

As "percent" it is used to describe the steepness of the slope of a road or railway, formula for which is 100 × rise/run which could also be expressed as the tangent of the angle of inclination times 100. This is the ratio of distances a vehicle would advance vertically and horizontally, respectively, when going up- or downhill, expressed in percent. Percentage is also used to express composition of a mixture by mass percent and mole percent.

4

Ratio and Proportion

Ratio and Proportion are explained majorly based on fractions. When a fraction is represented in the form of a:b, then it is a ratio whereas a proportion states that two ratios are equal. Here, a and b are any two integers. The ratio and proportion are the two important concepts, and it is the foundation to understand the various concepts in mathematics as well as in science.

In our daily life, we use the concept of ratio and proportion such as in business while dealing with money or while cooking any dish, etc. Sometimes, students get confused with the concept of ratio and proportion. In this article, the students get a clear vision of these two concepts with more solved examples and problems.

The definition of ratio and proportion is described here in this section. Both concepts are an important part of Mathematics. In real life also, you may find a lot of examples such as the rate of speed (distance/time) or price (rupees/meter) of a material, etc, where the concept of the ratio is highlighted.

Proportion is an equation that defines that the two given ratios are equivalent to each other. For example, the time taken by train to cover 100km per hour is equal to the time taken by it to cover the distance of 500km for 5 hours. Such as 100 km/hr = 500 km/5hrs.

Ratios are everywhere. They are one of the most important concepts in mathematics because of just how important they are in our day-to-day life. They play a major role in several things. For example, things like just comparing shoe sizes, or shirt sizes. Ratios are perhaps one of the most commonly used concepts that you definitely should know about. Even if you are new to the concept of ratio and proportions, chances are, nine out of ten times, you have heard of the two terms "ratio" and "proportions". We are surrounded by them. They are everywhere.

However, if you have not and you are completely new to the concept, then we are here to help you out with that. Ratio and proportions are simple to understand and can be practised easily to master so that the fundamentals that you learn can be used in your everyday life. This article majorly focuses on the concept of the ratio and proportion formula. It is a very well-written and detailed approach towards the formula that we suggest is going to help you a lot in understanding how things work in the world of ratios.

Ratio and proportions are an integral part of our day-to-day lives. Math is all around us because of topics and concepts like these. We hope that this article was successful in telling what exactly the study of ratio and proportion is all about and gave you a good explanation of the idea and concept of the ratio and proportion formula. We recommend that students use the examples given in the article as references every time they solve and use the formula. Thank you so much for reading and we hope that this article helped!

What do we know about the concept of ratios and proportions in general arithmetic? When we talk about the speed of a car and the distance it covers per km or maybe a recipe of a dish, we are talking about it in relation to ratios and proportions. But what actually are the concepts of ratio and proportions? What is the ratio and proportion formula? And why do we use the ratio and proportion formula? Why are they important to us? How do we use them in our day-to-day life? Let us elucidate all the clouds of questions one by one.

RATIO MEANING

We always compare two or more things in our everyday life as per the need. This comparison between two or more quantities using the method of division is extremely efficient. So, it will not be wrong if we say that ratio is actually, the comparison or simplification of two quantities of the same kind. This relation explains to us how many times is one quantity equal to the other quantity. In simple words, the ratio is the number that we use to express one quantity as a fraction of the other ones.

We can only compare two numbers in a ratio if they have the same unit and the sign that we use to denote a ratio is ":". In a fraction, it is written using "/" and we also use "to" to represent a ratio.

In certain situations, the comparison of two quantities by the method of division is very efficient. We can say that the comparison or simplified form of two quantities of the same kind is referred to as a ratio. This relation gives us how many times one quantity is equal to the other quantity. In simple words, the ratio is the number that can be used to express one quantity as a fraction of the other ones.

The two numbers in a ratio can only be compared when they have the same unit. We make use of ratios to compare two things. The sign used to denote a ratio is ':'.

A ratio can be written as a fraction, say 2/5. We happen to see various comparisons or say ratios in our daily life.

Hence, the ratio can be represented in three different forms, such as:

- a to b
- a : b
- a/b

Key Points to Remember:

- The ratio should exist between the quantities of the same kind
- While comparing two things, the units should be similar
- There should be significant order of terms
- The comparison of two ratios can be performed, if the ratios are equivalent like the fractions

Notation and terminology

The ratio of numbers A and B can be expressed as:

- the ratio of A to B
- A:B
- A is to B (when followed by "as C is to D "; see below)
- a fraction with A as numerator and B as denominator that represents the quotient (i.e., A divided by B, or $\frac{A}{B}$). This can be expressed as a simple or a decimal fraction, or as a percentage, etc.

A colon (:) is often used in place of the ratio symbol, Unicode U+2236 (:).

The numbers A and B are sometimes called terms of the ratio, with A being the antecedent and B being the consequent.

A statement expressing the equality of two ratios A:B and C:D is called a proportion, written as A:B = C:D or A:B7"C:D. This latter form, when spoken or written in the English language, is often expressed as

(A is to B) as (C is to D).

A, B, C and D are called the terms of the proportion. A and D are called its extremes, and B and C are called its means. The equality of three or more ratios, like A:B = C:D = E:F, is called a continued proportion.

Ratios are sometimes used with three or even more terms, e.g., the proportion for the edge lengths of a "two by four" that is ten inches long is therefore

$$\text{thickness} : \text{width} : \text{length} = 2 : 4 : 10;$$

(unplaned measurements; the first two numbers are reduced slightly when the wood is planed smooth)

a good concrete mix (in volume units) is sometimes quoted as

cement : sand : gravel = 1 : 2 : 4.

For a (rather dry) mixture of 4/1 parts in volume of cement to water, it could be said that the ratio of cement to water is 4:1, that there is 4 times as much cement as water, or that there is a quarter (1/4) as much water as cement.

The meaning of such a proportion of ratios with more than two terms is that the ratio of any two terms on the left-hand side is equal to the ratio of the corresponding two terms on the right-hand side.

History and etymology

It is possible to trace the origin of the word "ratio" to the Ancient Greek ëüãïò (logos). Early translators rendered this into Latin as ratio ("reason"; as in the word "rational"). A more modern interpretation[compared to?] of Euclid's meaning is more akin to computation or reckoning. Medieval writers used the word proportio ("proportion") to indicate ratio and proportionalitas ("proportionality") for the equality of ratios.

Euclid collected the results appearing in the Elements from earlier sources. The Pythagoreans developed a theory of ratio and proportion as applied to numbers. The Pythagoreans' conception of number included only what would today be called rational numbers, casting doubt on the validity of the theory in geometry where, as the Pythagoreans also discovered, incommensurable ratios (corresponding to irrational numbers) exist. The discovery of a theory of ratios that does not assume commensurability is probably due to Eudoxus of Cnidus. The exposition of the theory of proportions that appears in Book VII of The Elements reflects the earlier theory of ratios of commensurables.

The existence of multiple theories seems unnecessarily complex since ratios are, to a large extent, identified with quotients and their prospective values. However, this is a comparatively recent development, as can be seen from the fact that modern geometry textbooks still use distinct terminology and notation for ratios and quotients. The reasons for this are twofold: first, there was the previously mentioned reluctance to accept irrational numbers as true numbers, and second, the lack of a widely used symbolism to replace the already established terminology of ratios delayed the full acceptance of fractions as alternative until the 16th century.

Euclid's definitions

Book V of Euclid's Elements has 18 definitions, all of which relate to ratios. In addition, Euclid uses ideas that were in such common usage that he did not include definitions for them. The first two definitions say that a part of a quantity is another quantity that "measures" it and conversely, a multiple of a quantity is another quantity that it measures. In modern terminology, this means that a multiple of a quantity is that quantity multiplied by an integer greater than one—and a part of a quantity (meaning aliquot part) is a part that, when multiplied by an integer greater than one, gives the quantity.

Euclid does not define the term "measure" as used here, However, one may infer that if a quantity is taken as a unit of measurement, and a second quantity is given as an integral number of these units, then the first quantity measures the second. These definitions are repeated, nearly word for word, as definitions 3 and 5 in book VII.

Definition 3 describes what a ratio is in a general way. It is not rigorous in a mathematical sense and some have ascribed it to Euclid's editors rather than Euclid himself. Euclid defines a ratio as between two quantities of the same type, so by this definition the ratios of two lengths or of two areas are defined, but not the ratio of a length and an area. Definition 4 makes this more rigorous. It states that a ratio of two quantities exists, when there is a multiple of each that exceeds the other. In modern notation, a ratio exists between quantities p and q, if there exist integers m and n such that $mp>q$ and $nq>p$. This condition is known as the Archimedes property.

Definition 5 is the most complex and difficult. It defines what it means for two ratios to be equal. Today, this can be done by simply stating that ratios are equal when the quotients of the terms are equal, but such a definition would have been meaningless to Euclid. In modern notation, Euclid's definition of equality is that given quantities p, q, r and s, p:q7"r :s if and only if, for any positive integers m and n, $np<mq$, $np=mq$, or $np>mq$ according as $nr<ms$, $nr=ms$, or $nr>ms$, respectively. This definition has affinities with Dedekind cuts as, with n and q both positive, np stands to mq as p/q stands to the rational number m/n (dividing both terms by nq).

Definition 6 says that quantities that have the same ratio are proportional or in proportion. Euclid uses the Greek íáëüãïí (analogon), this has the same root as ëüãïò and is related to the English word "analog".

Definition 7 defines what it means for one ratio to be less than or greater than another and is based on the ideas present in definition 5. In modern notation it says that given quantities p, q, r and s, $p:q>r:s$ if there are positive integers m and n so that $np>mq$ and nrd"ms.

As with definition 3, definition 8 is regarded by some as being a later insertion by Euclid's editors. It defines three terms p, q and r to be in proportion when p:q7"q:r. This is extended to 4 terms p, q, r and s as p:q7"q:r7"r:s, and so on. Sequences that have the property that the ratios of consecutive terms are equal are called geometric progressions. Definitions 9 and 10 apply this, saying that if p, q and r are in proportion then p:r is the duplicate ratio of p:q and if p, q, r and s are in proportion then p:s is the triplicate ratio of p:q.

Number of terms and use of fractions

In general, a comparison of the quantities of a two-entity ratio can be expressed as a fraction derived from the ratio. For example, in a ratio of 2:3, the amount, size, volume, or quantity of the first entity is $\frac{2}{3}$ that of the second entity.

If there are 2 oranges and 3 apples, the ratio of oranges to apples is 2:3, and the ratio of oranges to the total number of pieces of fruit is 2:5. These ratios can also be expressed in fraction form: there are 2/3 as many oranges as apples, and 2/5 of the pieces of fruit are oranges. If orange juice concentrate is to be diluted with water in the ratio 1:4, then one part of concentrate is mixed with four parts of water, giving five parts total; the amount of orange juice concentrate is 1/4 the amount of water, while the amount of orange juice concentrate is 1/5 of the total liquid. In both ratios and fractions, it is important to be clear what is being compared to what, and beginners often make mistakes for this reason.

Fractions can also be inferred from ratios with more than two entities; however, a ratio with more than two entities cannot be completely converted into a single fraction, because a fraction can only compare two quantities. A separate fraction can be used to compare the quantities of any two of the entities covered by the ratio: for example, from a ratio of 2:3:7 we can infer that the quantity of the second entity is $\frac{3}{7}$ that of the third entity.

Proportions and percentage ratios

If we multiply all quantities involved in a ratio by the same number, the ratio remains valid. For example, a ratio of 3:2 is the same as 12:8. It is usual either to reduce terms to the lowest common denominator, or to express them in parts per hundred (percent).

If a mixture contains substances A, B, C and D in the ratio 5:9:4:2 then there are 5 parts of A for every 9 parts of B, 4 parts of C and 2 parts of D. As 5+9+4+2=20, the total mixture contains 5/20 of A (5 parts out of 20), 9/20 of B, 4/20 of C, and 2/20 of D. If we divide all numbers by the total and multiply by 100, we have converted to percentages: 25% A, 45% B, 20% C, and 10% D (equivalent to writing the ratio as 25:45:20:10).

If the two or more ratio quantities encompass all of the quantities in a particular situation, it is said that "the whole" contains the sum of the parts:

for example, a fruit basket containing two apples and three oranges and no other fruit is made up of two parts apples and three parts oranges. In this case, $\frac{2}{5}$, or 40% of the whole is apples and $\frac{3}{5}$, or 60% of the whole is oranges. This comparison of a specific quantity to "the whole" is called a proportion.

If the ratio consists of only two values, it can be represented as a fraction, in particular as a decimal fraction. For example, older televisions have a 4:3 aspect ratio, which means that the width is 4/3 of the height (this can also be expressed as 1.33:1 or just 1.33 rounded to two decimal places). More recent widescreen TVs have a 16:9 aspect ratio, or 1.78 rounded to two decimal places. One of the popular widescreen movie formats is 2.35:1 or simply 2.35. Representing ratios as decimal fractions simplifies their comparison. When comparing 1.33, 1.78 and 2.35, it is obvious which format offers wider image. Such a comparison works only when values being compared are consistent, like always expressing width in relation to height.

REDUCTION

Ratios can be reduced (as fractions are) by dividing each quantity by the common factors of all the quantities. As for fractions, the simplest form is considered that in which the numbers in the ratio are the smallest possible integers.

Thus, the ratio 40:60 is equivalent in meaning to the ratio 2:3, the latter being obtained from the former by dividing both quantities by 20. Mathematically, we write 40:60 = 2:3, or equivalently 40:607"2:3. The verbal equivalent is "40 is to 60 as 2 is to 3."

A ratio that has integers for both quantities and that cannot be reduced any further (using integers) is said to be in simplest form or lowest terms.

Sometimes it is useful to write a ratio in the form 1:x or x:1, where x is not necessarily an integer, to enable comparisons of different ratios. For example, the ratio 4:5 can be written as 1:1.25 (dividing both sides by 4) Alternatively, it can be written as 0.8:1 (dividing both sides by 5).

Where the context makes the meaning clear, a ratio in this form is sometimes written without the 1 and the ratio symbol (:), though, mathematically, this makes it a factor or multiplier.

Odds

Odds (as in gambling) are expressed as a ratio. For example, odds of "7 to 3 against" (7:3) mean that there are seven chances that the event will not happen to every three chances that it will happen. The probability of success is 30%. In every ten trials, there are expected to be three wins and seven losses.

Units

Ratios may be unitless, as in the case they relate quantities in units of the same dimension, even if their units of measurement are initially different. For example, the ratio 1 minute : 40 seconds can be reduced by changing the first value to 60 seconds, so the ratio becomes 60 seconds : 40 seconds. Once the units are the same, they can be omitted, and the ratio can be reduced to 3:2.

On the other hand, there are non-dimensionless ratios, also known as rates. In chemistry, mass concentration ratios are usually expressed as weight/volume fractions. For example, a concentration of 3% w/v usually means 3 g of substance in every 100 mL of solution. This cannot be converted to a dimensionless ratio, as in weight/weight or volume/volume fractions.

Triangular coordinates

The locations of points relative to a triangle with vertices A, B, and C and sides AB, BC, and CA are often expressed in extended ratio form as triangular coordinates.

In barycentric coordinates, a point with coordinates á, â, ã is the point upon which a weightless sheet of metal in the shape and size of the triangle would exactly balance if weights were put on the vertices, with the ratio of the weights at A and B being á : â, the ratio of the weights at B and C being â : ã, and therefore the ratio of weights at A and C being á : ã.

In trilinear coordinates, a point with coordinates x :y :z has perpendicular distances to side BC (across from vertex A) and side CA (across from vertex B) in the ratio x :y, distances to side CA and side AB (across from C) in the ratio y :z, and therefore distances to sides BC and AB in the ratio x :z.

Since all information is expressed in terms of ratios (the individual numbers denoted by á, â, ã, x, y, and z have no meaning by themselves), a triangle analysis using barycentric or trilinear coordinates applies regardless of the size of the triangle.

PROPORTIONS

In mathematics, the word "proportions" means two ratios put into an equation. Some examples of proportions are:

- $^{50}D_{100} = {}^{1}D_{2}$
- $^{75}D_{100} = {}^{3}D_{4}$
- $^{x}D_{100} = {}^{3}D_{4}$, where x = 75.

In algebra, proportions can be used to solve many common problems about changing numbers. As an example, for the increase in a $40 purchase of gasoline (petrol), if the price rose 35 cents, from $3.50 to $3.85, then the proportion would be:

- $^{x}D_{3.85} = {}^{\$40}D_{3.50}$

The solution is simply:

- x = \$40/3.50 x 3.85 = \$44.00, or \$4 more when \$0.35 higher.

Many other common calculations can be solved by using proportions to show the relationships between the numbers.

In statistics, a proportion is a number which measures the extent to which a specific characteristic is in a sample or a population. It can be thought of as a percentage. To represent a sample proportion, the letters p can be used. To represent a population proportion, the letter π can be used.

DEFINITION OF PROPORTION

Proportion is an equation that defines that the two given ratios are equivalent to each other. In other words, the proportion states the equality of the two fractions or the ratios. In proportion, if two sets of given numbers are increasing or decreasing in the same ratio, then the ratios are said to be directly proportional to each other.

For example, the time taken by train to cover 100km per hour is equal to the time taken by it to cover the distance of 500km for 5 hours. Such as 100km/hr = 500km/5hrs.

Ratio and proportions are said to be faces of the same coin. When two ratios are equal in value, then they are said to be in proportion. In simple words, it compares two ratios. Proportions are denoted by the symbol '::' or '='.

The proportion can be classified into the following categories, such as:

- Direct Proportion
- Inverse Proportion
- Continued Proportion

Now, let us discuss all these methods in brief:

Direct Proportion

The direct proportion describes the relationship between two quantities, in which the increases in one quantity, there is an increase in the other quantity also. Similarly, if one quantity decreases, the other quantity also decreases. Hence, if "a" and "b" are two quantities, then the direction proportion is written as a""b.

Inverse Proportion

The inverse proportion describes the relationship between two quantities in which an increase in one quantity leads to a decrease in the other quantity. Similarly, if there is a decrease in one quantity, there is an increase in the other quantity. Therefore, the inverse proportion of two quantities, say "a" and "b" is represented by a""(1/b).

Continued Proportion

Consider two ratios to be a: b and c: d.

Then in order to find the continued proportion for the two given ratio terms, we convert the means to a single term/number. This would, in general, be the LCM of means.

For the given ratio, the LCM of b & c will be bc.

Thus, multiplying the first ratio by c and the second ratio by b, we have

First ratio- ca:bc

Second ratio- bc: bd

Thus, the continued proportion can be written in the form of ca: bc: bd

Proportion validates if the two ratios are equivalent to each other. It judges the equality of two ratios. Now, for example, consider that two sets of numbers are given to you that are increasing or decreasing in the same ratio. So in proportion, we will say that the ratios are directly proportional to each other. Let us take another example, a train that is covering 100km per hour is actually equivalent to a train covering the distance of 500km for 5 hours because 100km/hr = 500km/5hrs = 100km/hr.

Two Faces of a Coin

I think we are now very clear about the concepts of ratio and proportion. Ratios and proportions are actually the two faces of the same coin. If two ratios are equal, then it is a proportion. Ratios and proportions are normally defined on the basis of fractions. When we define a fraction using ":", it becomes a ratio and when we represent two ratios using "::", it is a proportion. Both ratio and proportion are an important foundation that helps us to understand many crucial concepts in Maths and Science. The formula for ratio proportion is the foundation of many such concepts.

Proportionality constant

A proportionality constant is a number that is used to convert a measurement in one system to the equivalent measurement in another system. For instance, people who are familiar with the traditional system of units used in the United States (pounds, feet, inches, etc.) may need to find out the metric equivalent for these measures in grams and meters. To make these calculations, they would need some proportionality constants.

One way to write a formula showing how to use a proportionality constant K is:

$KX = Y$

For example, people may know that they have 100 eggs and want to know how many dozen eggs they have. The proportionality constant K is then 1 dozen/ 12 eggs.

100 eggs × (1 dozen / 12 eggs) = 8 dozen eggs + 4 eggs

In general, given two functions f(x) and g(x), if there is a constant K such that f(x) = Kg(x), then we say that "f is directly proportional to g". In symbols, this can be written as f(x) "" g (x).

PROPORTION FORMULA

Now, let us assume that, in proportion, the two ratios are a:b & c:d. The two terms 'b' and 'c' are called 'means or mean term,' whereas the terms 'a' and 'd' are known as 'extremes or extreme terms.'

a/b = c/d or a : b :: c : d

Example: Let us consider one more example of a number of students in a classroom. Our first ratio of the number of girls to boys is 3:5 and that of the other is 4:8, then the proportion can be written as:

3 : 5 :: 4 : 8 or 3/5 = 4/8

Here, 3 & 8 are the extremes, while 5 & 4 are the means.

Note: The ratio value does not affect when the same non-zero number is multiplied or divided on each term.

Important Properties of Proportion

The following are the important properties of proportion:

- Addendo – If a : b = c : d, then a + c : b + d
- Subtrahendo – If a : b = c : d, then a – c : b – d
- Dividendo – If a : b = c : d, then a – b : b = c – d : d
- Componendo – If a : b = c : d, then a + b : b = c+d : d
- Alternendo – If a : b = c : d, then a : c = b: d
- Invertendo – If a : b = c : d, then b : a = d : c
- Componendo and dividendo – If a : b = c : d, then a + b : a – b = c + d : c – d

Difference Between Ratio and Proportion

To understand the concept of ratio and proportion, go through the difference between ratio and proportion given here.

S.No	Ratio	Proportion
1	The ratio is used to compare the size of two things with the same unit	The proportion is used to express the relation of two ratios
2	It is expressed using a colon (:), slash (/)	It is expressed using the double colon (::) or equal to the symbol (=)
3	It is an expression	It is an equation
4	Keyword to identify ratio in a problem is "to every"	Keyword to identify proportion in a problem is "out of"

Fourth, Third and Mean Proportional

If a : b = c : d, then:

- d is called the fourth proportional to a, b, c.
- c is called the third proportion to a and b.
- Mean proportional between a and b is —(ab).

Comparison of Ratios

If (a:b)>(c:d) = (a/b>c/d)

The compounded ratio of the ratios: (a : b), (c : d), (e : f) is (ace : bdf).

Duplicate Ratios

If a:b is a ratio, then:

- a^2:b^2 is a duplicate ratio
- "a:"b is the sub-duplicate ratio
- a^3:b^3 is a triplicate ratio

Ratio and Proportion Tricks

Let us learn here some rules and tricks to solve problems based on ratio and proportion topics.

- If u/v = x/y, then uy = vx
- If u/v = x/y, then u/x = v/y
- If u/v = x/y, then v/u = y/x
- If u/v = x/y, then (u+v)/v = (x+y)/y
- If u/v = x/y, then (u-v)/v = (x-y)/y
- If u/v = x/y, then (u+v)/ (u-v) = (x+y)/(x-y), which is known as componendo -Dividendo Rule
- If a/(b+c) = b/(c+a) = c/(a+b) and a+b+ c 'Š0, then a =b = c

PERCENT AND PROPORTIONS

When the ratio of part to whole is equal to the ratio of a percent to 100, shows a percent equal to an equivalent ratio, we call it a percent proportion. A percentage is a fraction expressed with 100 as the denominator. When two ratios are equal, they are said to be in proportion. Proportions are denoted by using the symbols "=" or "::". For example, 1/5 :: 20/ 100, and the given proportion is read as "1 is to 5 equals to 20 is to 100".

Every statement of percent can be expressed verbally as: "One number is some percent of another number." Percent statements will always involve three numbers. For example:

____ is ____ % of ____.

In the problem, 8 is what percent of 20?, the number 8 is some percent of the number 20. Looking at this problem, it is clear that 8 is the part and 20 is the whole. Similarly, in the statement, "One number is some percent of another number.", the phrase "one number" represents the part and "another number" represents the whole. Thus the statement, "One number is some percent of another number.", can be rewritten:

"One number is some percent of another number.", becomes, "The part is some percent of the whole."

Percentage in the Form of a Fraction Shows Proportion

Following are the given examples that show how the percentage is converted into fractions and shows the proportion.

- 25% means 25 parts out of 100 and 25/100 = 1/4 of the total parts (whole). So, the given proportion is read as "25 is to 100 equals to 1 is to 4".
- 40% means 40 parts out of 100 and 40/100 = 2/5 of the total parts (whole). So, the given proportion is read as "40 is to 100 equals to 2 is to 5".
- 75% means 75 parts out of 100 and 75/100 = 3/4 of the total parts (whole). So, the given proportion is read as "75 is to 100 equals to 3 is to 4".

Percent Proportion Formula

The percent proportion formula helps in solving problems. It is expressed in form of PART/WHOLE = PERCENT/100, where,

- PERCENT is the number with a percent sign.
- PART is the number with the word IS.
- WHOLE is the number with the word OF.

For example, 30 is what percent of 60? Here IS and OF are used in question where IS is PART and OF is WHOLE. Now with the help of formula, Part/Whole = Percent/100 ⇒! 30 / 60 = Percent /100 ⇒! Percent = (30 / 60) × 100 = 50%

Taking another example, What number is 10 % of 50?

Part/Whole = Percent/100 ⇒! Part/50 = 10/100 ⇒! Part = 5. Therefore, 10% of 50 is 5.

Ratio and Proportion Summary

- Ratio defines the relationship between the quantities of two or more objects. It is used to compare the quantities of the same kind.
- If two or more ratios are equal, then it is said to be in proportion.
- The proportion can be represented in two different ways. Either it can be represented using an equal sign or by using a colon symbol.(i.e) a:b = c:d or a:b :: c:d
- If we multiply or divide each term of the ratio by the same number, it does not affect the ratio.

- For any three quantities, the quantities are said to be in continued proportion, if the ratio between the first and second quantity is equal to the ratio between the second and third quantity.
- For any four quantities, they are said to be in continued proportion, if the ratio between the first and second quantities is equal to the ratio between the third and fourth quantities

Ratio and Proportion Examples

Example 1:

Are the ratios 4:5 and 8:10 said to be in Proportion?

Solution:

4:5= 4/5 = 0.8 and 8: 10= 8/10= 0.8

Since both the ratios are equal, they are said to be in proportion.

Example 2:

Are the two ratios 8:10 and 7:10 in proportion?

Solution:

8:10= 8/10= 0.8 and 7:10= 7/10= 0.7

Since both the ratios are not equal, they are not in proportion.

Example 3:

Given ratio are-

a:b = 2:3

b:c = 5:2

c:d = 1:4

Find a: b: c.

Solution:

Multiplying the first ratio by 5, second by 3 and third by 6, we have

a:b = 10: 15

b:c = 15 : 6

c:d = 6 : 24

In the ratio's above, all the mean terms are equal, thus

a:b:c:d = 10:15:6:24

Example 4:

Check whether the following statements are true or false.

a] 12 : 18 = 28 : 56

b] 25 people : 130 people = 15kg : 78kg

Solution:

a] 12 : 18 = 28 : 56

The given statement is false.

12 : 18 = 12 / 18 = 2 / 3 = 2 : 3

28 : 56 = 28 / 56 = 1 / 2 = 1 : 2

They are unequal.

b] 25 persons : 130 persons = 15kg : 78kg

The given statement is true.

25 people : 130 people = 5: 26

15kg : 78kg = 5: 26

They are equal.

Example 5:

The earnings of Rohan is 12000 rupees every month and Anish is 191520 per year. If the monthly expenses of every person are around 9960 rupees. Find the ratio of the savings.

Solution:

Savings of Rohan per month = Rs (12000-9960) = Rs. 2040

Yearly income of Anish = Rs. 191520

Hence, the monthly income of Anish = Rs. 191520/12 = Rs. 15960.

So, the savings of Anish per month = Rs (15960 – 9960) = Rs. 6000

Thus, the ratio of savings of Rohan and Anish is Rs. 2040: Rs.6000 = 17: 50.

Example 6:

Twenty tons of iron is Rs. 600000 (six lakhs). What is the cost of 560 kilograms of iron?

Solution:

1 ton = 1000 kg

20 tons = 20000 kg

The cost of 20000 kg iron = Rs. 600000

The cost of 1 kg iron = Rs{600000}/ {20000}

= Rs. 30

The cost of 560 kg iron = Rs 30 × 560 = Rs 16800

Example 7:

The dimensions of the rectangular field are given. The length and breadth of the rectangular field are 50 meters and 15 meters. What is the ratio of the length and breadth of the field?

Solution:

Length of the rectangular field = 50 m

Breadth of the rectangular field = 15 m

Hence, the ratio of length to breadth = 50: 15

⇒! 50: 15 = 10: 3.

Thus, the ratio of length and breadth of the rectangular field is 10:3.

Example 8:

Obtain a ratio of 90 centimetres to 1.5 meters.

Solution:

The given two quantities are not in the same units.

Convert them into the same units.

1.5 m = 1.5 × 100 = 150 cm

Hence, the required ratio is 90 cm: 150 cm

⇒! 90: 150 = 3: 5

Therefore, the ratio of 90 centimetres to 1.5 meters is 3: 5.

Example 9:

There exists 45 people in an office. Out of which female employees are 25 and the remaining are male employees. Find the ratio of

a] The count of females to males.

b] The count of males to females.

Solution:

Count of females = 25

Total count of employees = 45

Count of males = 45 – 25 = 20

The ratio of the count of females to the count of males

= 25 : 20

= 5 : 4

The count of males to the count of females

= 20 : 25

= 4 : 5

Example 10:

Write two equivalent ratios of 6: 4.

Answer:

Given Ratio : 6: 4, which is equal to 6/4.

Multiplying or dividing the same numbers on both numerator and denominator, we will get the equivalent ratio.

⇒(6×2)/(4×2) = 12/8 = 12: 8

⇒(6÷2)/(4÷2) = 3/2 = 3: 2

Therefore, the two equivalent ratios of 6: 4 are 3: 2 and 12: 8

Example 11:

Out of the total students in a class, if the number of boys is 5 and the number of girls is 3, then find the ratio between girls and boys.

Solution:

The ratio between girls and boys can be written as 3:5 (Girls: Boys). The ratio can also be written in the form of factor like 3/5.

Example 12:

Two numbers are in the ratio 2 : 3. If the sum of numbers is 60, find the numbers.

Solution:

Given, 2/3 is the ratio of any two numbers.

Let the two numbers be 2x and 3x.

As per the given question, the sum of these two numbers = 60

So, $2x + 3x = 60$

$5x = 60$

$x = 12$

Hence, the two numbers are;

$2x = 2 \times 12 = 24$

and

$3x = 3 \times 12 = 36$

24 and 36 are the required numbers.

Maths ratio and proportion are used to solve many real-world problems. Register with BYJU'S – The Learning App and get solutions for many difficult questions in easy methodology and followed by the step-by-step procedure.

5

Fractions and Mixed Numbers

A fraction represents a part of a whole or, more generally, any number of equal parts. When spoken in everyday English, a fraction describes how many parts of a certain size there are, for example, one-half, eight-fifths, three-quarters. A common, vulgar, or simple fraction (examples: $\frac{1}{2}$ and $\frac{17}{3}$) consists of a numerator, displayed above a line (or before a slash like 1D 2), and a non-zero denominator, displayed below (or after) that line. Numerators and denominators are also used in fractions that are not common, including compound fractions, complex fractions, and mixed numerals.

In positive common fractions, the numerator and denominator are natural numbers. The numerator represents a number of equal parts, and the denominator indicates how many of those parts make up a unit or a whole. The denominator cannot be zero, because zero parts can never make up a whole. For example, in the fraction 3/4, the numerator 3 indicates that the fraction represents 3 equal parts, and the denominator 4 indicates that 4 parts make up a whole. The picture to the right illustrates 3/4 of a cake.

A common fraction is a numeral which represents a rational number. That same number can also be represented as a decimal, a percent, or with a negative exponent. For example, 0.01, 1%, and 10^{-2} are all equal to the fraction 1/100. An integer can be thought of as having an implicit denominator of one (for example, 7 equals 7/1).

Other uses for fractions are to represent ratios and division. Thus the fraction 3/4 can also be used to represent the ratio 3:4 (the ratio of the part to the whole), and the division 3 ÷ 4 (three divided by four). The non-zero denominator rule, which applies when representing a division as a fraction, is an example of the rule that division by zero is undefined.

We can also write negative fractions, which represent the opposite of a positive fraction. For example, if 1/2 represents a half dollar profit, then

"1/2 represents a half dollar loss. Because of the rules of division of signed numbers (which states in part that negative divided by positive is negative), "1/2, "1/2 and 1/"2 all represent the same fraction — negative one-half. And because a negative divided by a negative produces a positive, "1/"2 represents positive one-half.

In mathematics the set of all numbers that can be expressed in the form a/b, where a and b are integers and b is not zero, is called the set of rational numbers and is represented by the symbol Q, which stands for quotient. A number is a rational number precisely when it can be written in that form (i.e., as a common fraction). However, the word fraction can also be used to describe mathematical expressions that are not rational numbers. Examples of these usages include algebraic fractions (quotients of algebraic expressions), and expressions that contain irrational numbers, such as and π/4.

VOCABULARY

In a fraction, the number of equal parts being described is the numerator (from Latin numerâtor, "counter" or "numberer"), and the type or variety of the parts is the denominator (from Latin dçnôminâtor, "thing that names or designates"). As an example, the fraction 8/5 amounts to eight parts, each of which is of the type named "fifth". In terms of division, the numerator corresponds to the dividend, and the denominator corresponds to the divisor.

Informally, the numerator and denominator may be distinguished by placement alone, but in formal contexts they are usually separated by a fraction bar. The fraction bar may be horizontal (as in 1/3), oblique (as in 2/5), or diagonal (as in 4D 9). These marks are respectively known as the horizontal bar; the virgule, slash (US), or stroke (UK); and the fraction bar, solidus, or fraction slash.[n 1] In typography, fractions stacked vertically are also known as "en" or "nut fractions", and diagonal ones as "em" or "mutton fractions", based on whether a fraction with a single-digit numerator and denominator occupies the proportion of a narrow en square, or a wider em square. In traditional typefounding, a piece of type bearing a complete fraction (e.g. 1/2) was known as a "case fraction," while those representing only part of fraction were called "piece fractions."

The denominators of English fractions are generally expressed as ordinal numbers, in the plural if the numerator is not 1. (For example, 2/5 and 3/5 are both read as a number of "fifths".) Exceptions include the denominator 2, which is always read "half" or "halves", the denominator 4, which may be alternatively expressed as "quarter"/"quarters" or as "fourth"/ "fourths", and the denominator 100, which may be alternatively expressed as "hundredth"/"hundredths" or "percent". When the denominator is 1, it may be expressed in terms of "wholes" but is more commonly ignored, with the numerator read out as a whole number. For example, 3/1 may be described as "three wholes", or simply as "three". When the numerator is 1, it may be omitted (as in "a tenth" or "each quarter").

The entire fraction may be expressed as a single composition, in which case it is hyphenated, or as a number of fractions with a numerator of one, in which case they are not. (For example, "two-fifths" is the fraction 2/5 and "two fifths" is the same fraction understood as 2 instances of 1/5.) Fractions should always be hyphenated when used as adjectives. Alternatively, a fraction may be described by reading it out as the numerator "over" the denominator, with the denominator expressed as a cardinal number. (For example, 3/1 may also be expressed as "three over one".) The term "over" is used even in the case of solidus fractions, where the numbers are placed left and right of a slash mark. (For example, 1/2 may be read "one-half", "one half", or "one over two".) Fractions with large denominators that are not powers of ten are often rendered in this fashion (e.g., 1/117 as "one over one hundred seventeen"), while those with denominators divisible by ten are typically read in the normal ordinal fashion (e.g., 6/1000000 as "six-millionths", "six millionths", or "six one-millionths").

FORMS OF FRACTIONS

Simple, common, or vulgar fractions

A simple fraction (also known as a common fraction or vulgar fraction, where vulgar is Latin for "common") is a rational number written as a/b or $\frac{a}{b}$, where a and b are both integers. As with other fractions, the denominator (b) cannot be zero. Examples include $\frac{1}{2}$, $-\frac{8}{5}$, $\frac{-8}{5}$, and $\frac{8}{-5}$. The term was originally used to distinguish this type of fraction from the sexagesimal fraction used in astronomy.

Common fractions can be positive or negative, and they can be proper or improper. Compound fractions, complex fractions, mixed numerals, and decimals are not common fractions; though, unless irrational, they can be evaluated to a common fraction.

- A unit fraction is a common fraction with a numerator of 1 (e.g., $\frac{1}{7}$). Unit fractions can also be expressed using negative exponents, as in 2"1, which represents 1/2, and $2^{"2}$, which represents $1/(2^2)$ or 1/4.

- A dyadic fraction is a common fraction in which the denominator is a power of two, e.g. $\frac{1}{8}=\frac{1}{2^3}$.

In Unicode, precomposed fraction characters are in the Number Forms block.

Proper and improper fractions

Common fractions can be classified as either proper or improper. When the numerator and the denominator are both positive, the fraction is called proper if the numerator is less than the denominator, and improper otherwise. The concept of an "improper fraction" is a late development, with the terminology deriving from the fact that "fraction" means "a piece", so a proper fraction must be less than 1. This was explained in the 17th century textbook The Ground of Arts.

In general, a common fraction is said to be a proper fraction, if the absolute value of the fraction is strictly less than one—that is, if the fraction is greater than Š1 and less than 1. It is said to be an improper fraction, or sometimes top-heavy fraction, if the absolute value of the fraction is greater than or equal to 1. Examples of proper fractions are 2/3, Š3/4, and 4/9, whereas examples of improper fractions are 9/4, Š4/3, and 3/3.

Reciprocals and the "invisible denominator"

The reciprocal of a fraction is another fraction with the numerator and denominator exchanged. The reciprocal of $\frac{3}{7}$, for instance, is $\frac{7}{3}$. The product of a fraction and its reciprocal is 1, hence the reciprocal is the multiplicative inverse of a fraction. The reciprocal of a proper fraction is improper, and the reciprocal of an improper fraction not equal to 1 (that is, numerator and denominator are not equal) is a proper fraction.

When the numerator and denominator of a fraction are equal (for example, $\frac{7}{7}$), its value is 1, and the fraction therefore is improper. Its reciprocal is identical and hence also equal to 1 and improper.

Any integer can be written as a fraction with the number one as denominator. For example, 17 can be written as $\frac{17}{1}$ where 1 is sometimes referred to as the invisible denominator. Therefore, every fraction or integer, except for zero, has a reciprocal. For example. the reciprocal of 17 is $\frac{1}{17}$.

Ratios

A ratio is a relationship between two or more numbers that can be sometimes expressed as a fraction. Typically, a number of items are grouped and compared in a ratio, specifying numerically the relationship between each group. Ratios are expressed as "group 1 to group 2 ... to group n". For example, if a car lot had 12 vehicles, of which

- 2 are white,
- 6 are red, and
- 4 are yellow,

then the ratio of red to white to yellow cars is 6 to 2 to 4. The ratio of yellow cars to white cars is 4 to 2 and may be expressed as 4:2 or 2:1.

A ratio is often converted to a fraction when it is expressed as a ratio to the whole. In the above example, the ratio of yellow cars to all the cars on the lot is 4:12 or 1:3. We can convert these ratios to a fraction, and say that 4/12 of the cars or 1/3 of the cars in the lot are yellow. Therefore, if a person randomly chose one car on the lot, then there is a one in three chance or probability that it would be yellow.

DECIMAL FRACTIONS AND PERCENTAGES

A decimal fraction is a fraction whose denominator is not given explicitly, but is understood to be an integer power of ten. Decimal fractions are commonly expressed using decimal notation in which the implied denominator is determined by the number of digits to the right of a decimal separator, the appearance of which (e.g., a period, a raised period (•), a comma) depends on the locale (for examples, see decimal separator). Thus, for 0.75 the numerator is 75 and the implied denominator is 10 to the second power, viz. 100, because there are two digits to the right of the decimal separator. In decimal numbers greater than 1 (such as 3.75), the fractional part of the number is expressed by the digits to the right of the decimal (with a value of 0.75 in this case). 3.75 can be written either as an improper fraction, 375/100, or as a mixed number, $3\frac{75}{100}$.

Decimal fractions can also be expressed using scientific notation with negative exponents, such as 6.023×10^{-7}, which represents 0.0000006023. The 10^{-7} represents a denominator of 10^7. Dividing by 10^7 moves the decimal point 7 places to the left.

Decimal fractions with infinitely many digits to the right of the decimal separator represent an infinite series. For example, 1/3 = 0.333... represents the infinite series 3/10 + 3/100 + 3/1000 + ...

Another kind of fraction is the percentage (Latin per centum meaning "per hundred", represented by the symbol %), in which the implied denominator is always 100. Thus, 51% means 51/100. Percentages greater than 100 or less than zero are treated in the same way, e.g. 311% equals 311/100, and "27% equals "27/100.

The related concept of permille or parts per thousand (ppt) has an implied denominator of 1000, while the more general parts-per notation, as in 75 parts per million (ppm), means that the proportion is 75/1,000,000.

Whether common fractions or decimal fractions are used is often a matter of taste and context. Common fractions are used most often when the denominator is relatively small. By mental calculation, it is easier to multiply 16 by 3/16 than to do the same calculation using the fraction's decimal equivalent (0.1875). And it is more accurate to multiply 15 by 1/3, for example, than it is to multiply 15 by any decimal approximation of one third. Monetary values are commonly expressed as decimal fractions with denominator 100, i.e., with two decimals, for example $3.75. However, as noted above, in pre-decimal British currency, shillings and pence were often given the form (but not the meaning) of a fraction, as, for example 3/6 (read "three and six") meaning 3 shillings and 6 pence, and having no relationship to the fraction 3/6.

Mixed numbers

A mixed numeral (also called a mixed fraction or mixed number) is a traditional denotation of the sum of a non-zero integer and a proper fraction (having the same sign). It is used primarily in measurement: $2\tfrac{3}{16}$ inches, for example. Scientific measurements almost invariably use decimal notation rather than mixed numbers. The sum can be implied without the use of a visible operator such as the appropriate "+". For example, in referring to two entire cakes and three quarters of another cake, the numerals denoting the integer part and the fractional part of the cakes can be written next to each other as $2\tfrac{3}{4}$ instead of the unambiguous notation $2+\tfrac{3}{4}$.

Negative mixed numerals, as in $-2\tfrac{3}{4}$. are treated like $-\left(2+\dfrac{3}{4}\right)\cdot$ Any such sum of a whole plus a part can be converted to an improper fraction by applying the rules of adding unlike quantities.

This tradition is, formally, in conflict with the notation in algebra where adjacent symbols, without an explicit infix operator, denote a product. In the expression 2x, the "understood" operation is multiplication. If x is replaced by, for example, the fraction 3/4, the "understood" multiplication needs to be replaced by explicit multiplication, to avoid the appearance of a mixed number.

When multiplication is intended, $2\tfrac{b}{c}$ may be written as

$$2\cdot\frac{b}{c},\ \text{or}\ 2\times\frac{b}{c},\quad \text{or}\ 2\left(\frac{b}{c}\right),\ldots$$

An improper fraction can be converted to a mixed number as follows:

1. Using Euclidean division (division with remainder), divide the numerator by the denominator. In the example, $\frac{11}{4}$, divide 11 by 4. 11 ÷ 4 = 2 remainder 3.
2. The quotient (without the remainder) becomes the whole number part of the mixed number. The remainder becomes the numerator of the fractional part. In the example, 2 is the whole number part and 3 is the numerator of the fractional part.
3. The new denominator is the same as the denominator of the improper fraction. In the example, it is 4. Thus, $\frac{11}{4} = 2\frac{3}{4}$.

ARITHMETIC WITH FRACTIONS

Like whole numbers, fractions obey the commutative, associative, and distributive laws, and the rule against division by zero.

Equivalent fractions

Equivalent fractions can be defined as fractions that may have different numerators and denominators but they represent the same value. For example, 9/12 and 6/8 are equivalent fractions because both are equal to 3/4 when simplified.

All equivalent fractions get reduced to the same fraction in their simplest form as seen in the above example. Explore the given lesson to get a better idea of how to find equivalent fractions and how to check if the given fractions are equivalent.

Multiplying the numerator and denominator of a fraction by the same (non-zero) number results in a fraction that is equivalent to the original fraction. This is true because for any non-zero number n, the fraction $\frac{n}{n}$ equals 1. Therefore, multiplying by $\frac{n}{n}$ is the same as multiplying by one, and any number multiplied by one has the same value as the original number. By way of an example, start with the fraction $\frac{1}{2}$. When the numerator and denominator are both multiplied by 2, the result is 2/4, which has the same value (0.5) as ½. To picture this visually, imagine cutting a cake into four pieces; two of the pieces together (2/4) make up half the cake (1/2).

Two or more fractions are said to be equivalent if they are equal to the same fraction when simplified. For example, the equivalent fractions of 1/ 5 are 5/25, 6/30, and 4/20, which on simplification, result in the same fraction, that is, 1/5.

Equivalent Fractions Definition

Equivalent fractions are defined as those fractions which are equal to the same value irrespective of their numerators and denominators. For example, both 6/12 and 4/8 are equal to 1/2, when simplified, which means they are equivalent in nature.

Facts About the Equivalent fractions

Equivalent fractions represents the same amount of distance or points on a number line to one another.

All equivalent fractions are reduced to the same fractions in their simplest forms by dividing both numerator and denominator by their greatest common factor.

How to Find Equivalent Fractions:

By multiplying the numerator and denominator of a fraction by the same non-zero number, we can change the fraction into equivalent of the original fraction. This is true because for any non-zero number a, the fraction a/a = 1. n n = 1 Therefore, multiplying by a/a is equivalent to multiplying by one, and any number multiplied by one has the same value as the original number.

For Example:

To find the equivalent fraction of 2/5we multiply both the numerator and denominator by 2, then we get equivalent fraction 4/10.

How to Simplify the Equivalent Fractions?

Dividing an equivalent fractions numerator and denominator by the same non-zero number will also yields an equivalent fraction. This is called simplifying or reducing the fraction. A simple fraction in which the numerator and denominator both are prime number is said to be irreducible.

For Example:

To find the equivalent fraction of 6/18 we divide both the numerator and denominator by their greatest common factor, then we get equivalent fraction 1/3.

How to Check Two Fractions are Equivalent Fraction:

To check whether the fractions are equivalent or not, just simplify all the fractions. If they reduce into the same fraction, then the fractions are equivalent fractions.

The fractions ½ and S! are not the same, hence the two fractions are not Equivalent fractions.

ADDITION AND SUBTRACTION

Equivalent fractions are an important tool when adding or subtracting fractions with different denominators. Let's look at an example:

John bought one half of a cake. He didn't know that while he was out, his wife, Linda, bought one fourth of a cake. When they got home, how much cake did they have altogether?

We are adding fractions here:

When we are adding or subtracting fractions, they must have the same denominator. But in the above case, the denominator are two and four (2 and 4). As we read earlier by multiplying the numerator and denominator of a fraction, we get an equivalent fraction.

How to Find Equivalent Fractions?

Equivalent fractions can be written by multiplying or dividing both the numerator and the denominator by the same number. This is the reason why these fractions get reduced to the same number when they are simplified. Let us understand the two ways in which we can make equivalent fractions:

- Multiply the numerator and the denominator by the same number.
- Divide the numerator and the denominator by the same number.

Multiply the Numerator and Denominator by the Same Number

To find the equivalent fractions for any given fraction, multiply the numerator and the denominator by the same number. For example, to find an equivalent fraction of 3/4, multiply the numerator 3 and the denominator 4 by the same number, say, 2. Thus, 6/8 is an equivalent fraction of 3/4. We can find some other equivalent fractions by multiplying the numerator and the denominator of the given fraction by the same number.

- 3/4= 3×34×3
  - =9/12
  - 3/4=3×44×4
  - =12/16
  - 3/4=3×54×5
- =15/20

Thus, the equivalent fractions of 3/4 are 6/8, 9/12, 12/16, and 15/20.

Divide the Numerator and Denominator by the Same Number

To find the equivalent fractions for any given fraction, divide the numerator and the denominator by the same number. For example, to find an equivalent fraction of 72/108, we will first find their common factors. We know that 2 is a common factor of both 72 and 108. Hence, an equivalent fraction of 72/108 can be found by dividing its numerator and denominator by 2. Thus, 36/54 is an equivalent fraction of 72/108. Let us see how the fraction is further simplified:

- 2 is a common factor of 36 and 54. Thus, 36/54= 36÷254÷2
  - = 18/27
  - Again, 3 is a common factor of 18 and 27. Thus, 18/27= 18÷327÷3
  - = 6/9
  - Again, 3 is a common factor of 6 and 9. Thus, 6/9=6÷39÷3
- = 2/3

Therefore, a few equivalent fractions of 72/108 are 36/54, 18/27, 6/9, and 2/3. Here, 2/3 is the simplified form of 72/108 as there is no common factor (other than 1) of 2 and 3.

How to Determine if Two Fractions are Equivalent?

We need to simplify the given fractions to find whether they are equivalent or not. Simplification to get equivalent numbers can be done to a point where both the numerator and denominator should still be whole numbers. There are various methods to identify if the given fractions are equivalent. Some of them are:

- Making the denominators the same.
- Finding the decimal form of both the fractions.
- Cross multiplication method.
- Visual method.

Let us identify whether 2/6 and 3/9 are equivalent fractions by each of these methods.

MAKING THE DENOMINATORS THE SAME

The denominators of the fractions, 2/6 and 3/9 are 6 and 9. The Least Common Multiple (LCM) of the denominators 6 and 9 is 18. Let us make the denominators of both fractions 18, by multiplying them with suitable numbers.

- 2/6=2×36×3
  - = 6/18
  - 3/9=3×29×2
- = 6/18

We can observe that both the fractions are equivalent to the same fraction 6/18. Thus, the given fractions are equivalent.

Note: If the fractions are NOT equivalent, we can check the greater or smaller fraction by looking at the numerator of both the resultant fractions. Hence, this method can also be used for comparing fractions.

Finding the Decimal form of Both the Fractions

Let us find the decimal form of both the fractions 2/6 and 3/9 to see if they give the same value.

- 2/6= 0.3333333...
- 3/9= 0.3333333...

The decimal values of both the fractions are the same and hence, they are equivalent.

SIMPLIFYING (REDUCING) FRACTIONS

Dividing the numerator and denominator of a fraction by the same non-zero number yields an equivalent fraction: if the numerator and the denominator of a fraction are both divisible by a number (called a factor) greater than 1, then the fraction can be reduced to an equivalent fraction with a smaller numerator and a smaller denominator. For example, if both the numerator and the denominator of the fraction a/b are divisible by c, then they can be written as a = cd and b = ce, and the fraction becomes $\frac{cd}{ce}$, which can be reduced by dividing both the numerator and denominator by c to give the reduced fraction d/e.

If one takes for c the greatest common divisor of the numerator and the denominator, one gets the equivalent fraction whose numerator and denominator have the lowest absolute values. One says that the fraction has been reduced to its lowest terms.

If the numerator and the denominator do not share any factor greater than 1, the fraction is already reduced to its lowest terms, and it is said to be irreducible, reduced, or in simplest terms. For example, 3/9 is not in lowest terms because both 3 and 9 can be exactly divided by 3. In contrast, 3/8 is in lowest terms—the only positive integer that goes into both 3 and 8 evenly is 1.

Using these rules, we can show that $\frac{5}{10} = \frac{1}{2} = \frac{10}{20} = \frac{50}{100}$, for example.

As another example, since the greatest common divisor of 63 and 462 is 21, the fraction $\frac{63}{462}$ can be reduced to lowest terms by dividing the numerator and denominator by 21:

$$\frac{63}{462} = \frac{63 \div 21}{462 \div 21} = \frac{3}{22}$$

The Euclidean algorithm gives a method for finding the greatest common divisor of any two integers.

Comparing fractions

Comparing fractions with the same positive denominator yields the same result as comparing the numerators:

$\frac{3}{4} > \frac{2}{4}$ because $3 > 2$, and the equal denominators 4 are positive.

If the equal denominators are negative, then the opposite result of comparing the numerators holds for the fractions:

$$\frac{3}{-4} < \frac{2}{-4} \text{ because } \frac{a}{-b} = \frac{-a}{b} \text{ and } -3 < -2.$$

If two positive fractions have the same numerator, then the fraction with the smaller denominator is the larger number. When a whole is divided into equal pieces, if fewer equal pieces are needed to make up the whole, then each piece must be larger. When two positive fractions have the same numerator, they represent the same number of parts, but in the fraction with the smaller denominator, the parts are larger.

One way to compare fractions with different numerators and denominators is to find a common denominator. To compare a/b and c/d, these are converted to a Å″ d b Å″ d and b Å″ c b Å″ d (where the dot signifies multiplication and is an alternative symbol to ×). Then bd is a common denominator and the numerators ad and bc can be compared. It is not necessary to determine the value of the common denominator to compare fractions – one can just compare ad and bc, without evaluating bd, e.g., comparing 2/3 ? 1/2 gives 4/6 > 3/6.

Because every negative number, including negative fractions, is less than zero, and every positive number, including positive fractions, is greater than zero, it follows that any negative fraction is less than any positive fraction. This allows, together with the above rules, to compare all possible fractions.

Addition

The first rule of addition is that only like quantities can be added; for example, various quantities of quarters. Unlike quantities, such as adding thirds to quarters, must first be converted to like quantities as described below: Imagine a pocket containing two quarters, and another pocket containing three quarters; in total, there are five quarters. Since four quarters is equivalent to one (dollar), this can be represented as follows:

$$\frac{2}{4} + \frac{3}{4} = \frac{5}{4} = 1\frac{1}{4}.$$

Adding unlike quantities

To add fractions containing unlike quantities (e.g. quarters and thirds), it is necessary to convert all amounts to like quantities. It is easy to work out the chosen type of fraction to convert to; simply multiply together the two denominators (bottom number) of each fraction. In case of an integer number apply the invisible denominator 1.

For adding quarters to thirds, both types of fraction are converted to twelfths, thus:

$$\frac{1}{4}+\frac{1}{3}=\frac{1\times 3}{4\times 3}+\frac{1\times 4}{3\times 4}=\frac{3}{12}+\frac{4}{12}=\frac{7}{12}.$$

Converting between decimals and fractions

To change a common fraction to a decimal, do a long division of the decimal representations of the numerator by the denominator (this is idiomatically also phrased as "divide the denominator into the numerator"), and round the answer to the desired accuracy. For example, to change 1/4 to a decimal, divide 1.00 by 4 ("4 into 1.00"), to obtain 0.25. To change 1/3 to a decimal, divide 1.000... by 3 ("3 into 1.000..."), and stop when the desired accuracy is obtained, e.g., at 4 decimals with 0.3333. The fraction 1/4 can be written exactly with two decimal digits, while the fraction 1/3 cannot be written exactly as a decimal with a finite number of digits. To change a decimal to a fraction, write in the denominator a 1 followed by as many zeroes as there are digits to the right of the decimal point, and write in the numerator all the digits of the original decimal, just omitting the decimal point. Thus $12.3456=\frac{123456}{10000}$.

Converting repeating decimals to fractions

Decimal numbers, while arguably more useful to work with when performing calculations, sometimes lack the precision that common fractions have. Sometimes an infinite repeating decimal is required to reach the same precision. Thus, it is often useful to convert repeating decimals into fractions.

A conventional way to indicate a repeating decimal is to place a bar (known as a vinculum) over the digits that repeat, for example 0.789 = 0.789789789... For repeating patterns that begin immediately after the decimal point, the result of the conversion is the fraction with the pattern as a numerator, and the same number of nines as a denominator. For example:

0.5 = 5/9

0.62 = 62/99

0.264 = 264/999

0.6291 = 6291/9999

If leading zeros precede the pattern, the nines are suffixed by the same number of trailing zeros:

0.05 = 5/90

0.000392 = 392/999000

0.0012 = 12/9900

If a non-repeating set of decimals precede the pattern (such as 0.1523987), one may write the number as the sum of the non-repeating and repeating parts, respectively:

0.1523 + 0.0000987

Then, convert both parts to fractions, and add them using the methods described above:

1523 / 10000 + 987 / 9990000 = 1522464 / 9990000

Alternatively, algebra can be used, such as below:

1. Let x = the repeating decimal:

x = 0.1523987

2. Multiply both sides by the power of 10 just great enough (in this case 10^4) to move the decimal point just before the repeating part of the decimal number:

10,000x = 1,523.987

3. Multiply both sides by the power of 10 (in this case 10^3) that is the same as the number of places that repeat:

10,000,000x = 1,523,987.987

4. Subtract the two equations from each other (if a = b and c = d, then a “ c = b “ d):

10,000,000x “ 10,000x = 1,523,987.987 “ 1,523.987

5. Continue the subtraction operation to clear the repeating decimal:

9,990,000x = 1,523,987 “ 1,523

= 1,522,464

6. Divide both sides by 9,990,000 to represent x as a fraction

x = 1522464/9990000

Multiplication of Fractions

Multiplying fractions starts with the multiplication of the given numerators, followed by the multiplication of the denominators. Then, the resultant fraction is simplified further and reduced to its lowest terms, if needed. Learn all about multiplying fractions in this article.

The multiplication of fractions is not like the addition or subtraction of fractions, where the denominator should be the same. Here, any two fractions with different denominators can easily be multiplied. The only thing to be kept in mind is that the fractions should not be in the mixed form, they should either be proper fractions or improper fractions. Let us learn how to multiply fractions through the following steps:

- Step 1: Multiply the numerators.
- Step 2: Multiply the denominators.
- Step 3: Reduce the resultant fraction to its lowest terms.

For example, let us multiply the following fractions: 1/3 × 3/5. We start by multiplying the numerators: 1 × 3 = 3, then, the denominators: 3 × 5 = 15. This can be written as: (1 × 3)/(3 × 5) = 3/15. Now, reduce this value to its lowest form. 3 is the greatest common factor of 3 and 15, so, divide both 3 and 15 by 3 to simplify the fraction. Therefore, 1/3 × 3/5 = 1/5.

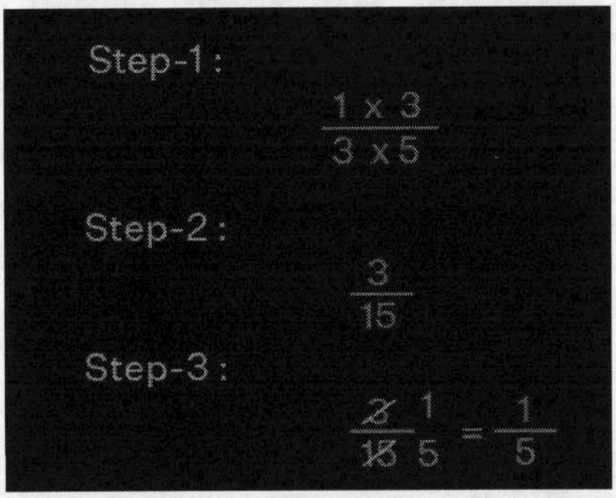

Multiplication of Fractions Using Visual Models

Now, let us see the visual representation for the multiplication of fractions. Visualizing the multiplication of fractions using fractional squares is a very interesting method to understand the concept. Let's multiply these two fractions: 1/4 × 1/3 using the visual model. Draw a rectangle and divide its length into 4 equal parts. Each column will represent 1/4 of the whole rectangle. Now, divide its width into 3 equal parts such that each portion will represent 1/3. Now, we just have to look for the portion that is common to both 1/4 and 1/3, which is 1/12th of the whole rectangle (marked in light orange color in the figure below).

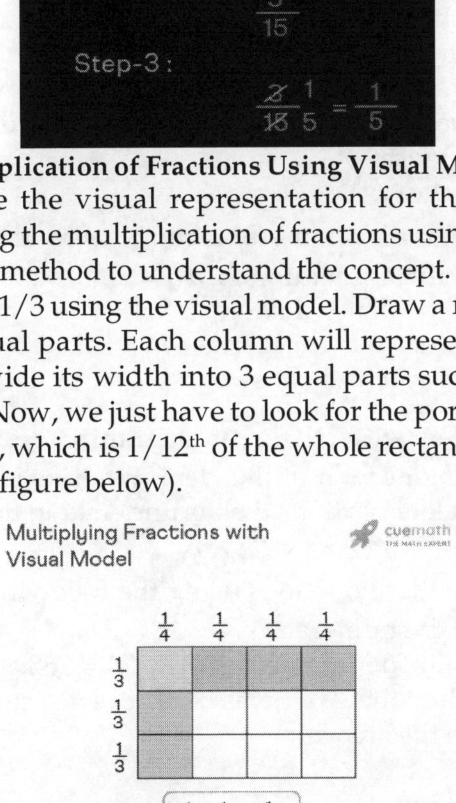

In the figure given above, we can clearly see that 1/12 of the rectangle is common to both 1/4 and 1/3. It is an overlapping portion. Thus, 1/4 ×

1/3 = 1/12. Now that you have insight into multiplying fractions, let us explore this topic further.

Rules of Multiplying Fractions

While multiplying fractions, the following rules have to be kept in mind:

- Rule 1: The first rule is to convert mixed fractions to improper fractions if any. Then, multiply the numerators of the given fractions.
- Rule 2: Multiply the denominators separately.
- Rule 3: Simplify the value obtained to its lowest term.

These three rules can be applied to any two fractions to find their product. Now, let us learn the individual cases of multiplying fractions with different types of fractions.

Multiplying Fractions with Same Denominator

Multiplying fractions with the same denominator does not change the rule of multiplication of fractions. Fractions that have the same denominator are termed like fractions. Although addition and subtraction of like fractions are different from the addition and subtraction of unlike fractions, in the case of multiplication and division the method remains the same. We multiply the numerators, then the denominators, and then the fraction is reduced to its lowest terms.

Example: Multiply 1/3 × 5/3

Solution: We can multiply these fractions using the following steps.

- Step 1: Multiply the numerators, 1 × 5 = 5.
- Step 2: Multiply the denominators, 3 × 3 = 9.
- Step 3: The product that we get is 5/9. This cannot be reduced any further, therefore 5/9 is the answer.

Multiplying Fractions with Different Denominators

Multiplying fractions with unlike denominators is exactly the same as the multiplication of like fractions. Let us understand this with an example.

Example: Multiply 4/12 × 16/24.

We can multiply these fractions using the following steps:

- Step 1: Multiply the numerators, 4 × 16 = 64.
- Step 2: Multiply the denominators, 12 × 24 = 288.
- Step 3: The product that we get is 64/288. This can be reduced to 2/9. Therefore, 2/9 is the answer.

Alternative Method

The same fractions can be multiplied using another method in which we simplify the fractions among themselves and then multiply the numerators, then the denominators to get the final product.

Example: Multiply 4/12 × 16/24.

Let us multiply the given fractions using the following steps:

- Step 1: We will simplify the given fractions among themselves. In other words, these fractions can be reduced to 1/3 × 2/3.
- Step 2: Let us multiply the numerators, 1 × 2 = 2.
- Step 3: Now, let us multiply the denominators, 3 × 3 = 9.
- Step 4: Therefore, the product that we get is 2/9.

Multiplying Fractions with Whole Numbers

Multiplying fractions by whole numbers is an easy concept. As we know that multiplication is the repeated addition of the same number, this fact can be applied to fractions as well.

Multiplying Fractions with Whole Numbers Visual Model

Let us consider this example: 4 × 2/3. This means 2/3 is added 4 times. Let us represent this example using a visual model. Four times two-thirds is represented as:

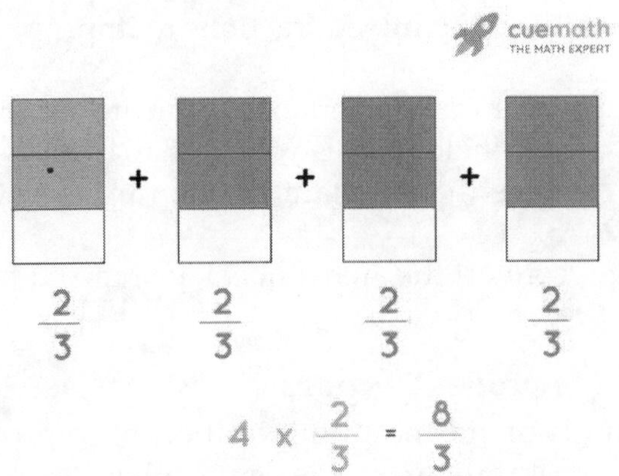

Steps of Multiplying Fractions with Whole Numbers

In order to multiply fractions with whole numbers, we use the simple rule of multiplying the numerators, then multiplying the denominators, and then reducing them to the lowest terms. However, in the case of whole numbers, we write them in the fractional form by placing '1' in the denominator. Let us understand this with an example.

Example: Multiply: 5 × 3/4.

Let us use the following steps to multiply the given fraction with a whole number.

- Step 1: Here, 5 is a whole number that can be written as 5/1, and then it can be multiplied as we multiply regular fractions.
- Step 2: This means, we need to multiply 5/1 × 3/4.

- Step 3: Multiply the numerators, 5 × 3 = 15.
- Step 4: Multiply the denominators, 1 × 4 = 4.
- Step 5: The resultant product is 15/4 which cannot be reduced further.
- Step 6: Since 15/4 is an improper fraction, we will change it to a mixed fraction, 15/4 = 334.

Multiplying Fractions with Mixed Numbers

Mixed numbers or mixed fractions are fractions that consist of a whole number and a proper fraction, like 234, where 2 is the whole number and 3/4 is the proper fraction. For multiplying mixed fractions, we need to change the mixed fractions into an improper fraction before multiplying. For example, if the number is 223, we need to change this to 8/3. Let us understand this with the help of an example.

Example: Multiply 223 and 314

The following steps can be used to multiply fractions with mixed numbers.

- Step 1: Change the given mixed fractions to improper fractions, i.e. (8/3) × (13/4).
- Step 2: Multiply the numerators of the improper fractions, and then multiply the denominators. This will give 104/12.
- Step 3: Now, reduce the resultant fraction to its lowest terms, which will make it 26/3.
- Step 4: Further, convert the answer back to a mixed fraction which will be 823.

Multiplication of Improper Fractions

Now let us understand the multiplication of improper fractions. We already know that an improper fraction is one where the numerator is bigger than the denominator. When multiplying two improper fractions, we frequently end up with an improper fraction. For example, to multiply 3/2 × 7/5 which are two improper fractions, we need to take the following steps:

- Step 1: Multiply the numerators and denominators. (3 × 7)/(2 × 5) = 21/10.
- Step 2: The fraction 21/10 cannot be reduced further to its lowest terms.
- Step 3: Hence, the answer is 21/10 which can be written as 2110.

Tips and Tricks of Multiplying Fractions:

Here are a few important tips and tricks which are helpful in the multiplication of fractions.

- Generally, students simplify a fraction after multiplication. However, to make calculations easier, check if the two fractions to be multiplied are already in their lowest forms. If not, first simplify them and then multiply. For example, 4/12 × 5/13 will be difficult to multiply directly.

- Now, if we simplify the fraction first, we get:1/3 × 5/13 = 5/39.
- Simplification can also be done across two fractions. If there is a common factor between the numerator of one of the fractions and the denominator of the other fraction, you can simplify them and proceed. For example, 5/28 × 7/9 can be simplified to 5/4 × 1/9 before multiplying.

Typographical variations

In computer displays and typography, simple fractions are sometimes printed as a single character, e.g. ½ (one half). See the article on Number Forms for information on doing this in Unicode.

Scientific publishing distinguishes four ways to set fractions, together with guidelines on use:

- special fractions: fractions that are presented as a single character with a slanted bar, with roughly the same height and width as other characters in the text. Generally used for simple fractions, such as: ½, S!, T!, ¼, and ¾. Since the numerals are smaller, legibility can be an issue, especially for small-sized fonts. These are not used in modern mathematical notation, but in other contexts.
- case fractions: similar to special fractions, these are rendered as a single typographical character, but with a horizontal bar, thus making them upright. An example would be $\frac{1}{2}$, but rendered with the same height as other characters. Some sources include all rendering of fractions as case fractions if they take only one typographical space, regardless of the direction of the bar.
- shilling or solidus fractions: 1/2, so called because this notation was used for pre-decimal British currency (£sd), as in 2/6 for a half crown, meaning two shillings and six pence. While the notation "two shillings and six pence" did not represent a fraction, the forward slash is now used in fractions, especially for fractions inline with prose (rather than displayed), to avoid uneven lines. It is also used for fractions within fractions (complex fractions) or within exponents to increase legibility. Fractions written this way, also known as piece fractions, are written all on one typographical line, but take 3 or more typographical spaces.
- built-up fractions: $\frac{1}{2}$. This notation uses two or more lines of ordinary text, and results in a variation in spacing between lines when included

within other text. While large and legible, these can be disruptive, particularly for simple fractions or within complex fractions.

History

The earliest fractions were reciprocals of integers: ancient symbols representing one part of two, one part of three, one part of four, and so on. The Egyptians used Egyptian fractions c. 1000 BC. About 4000 years ago, Egyptians divided with fractions using slightly different methods. They used least common multiples with unit fractions. Their methods gave the same answer as modern methods. The Egyptians also had a different notation for dyadic fractions in the Akhmim Wooden Tablet and several Rhind Mathematical Papyrus problems.

The Greeks used unit fractions and (later) continued fractions. Followers of the Greek philosopher Pythagoras (c. 530 BC) discovered that the square root of two cannot be expressed as a fraction of integers. (This is commonly though probably erroneously ascribed to Hippasus of Metapontum, who is said to have been executed for revealing this fact.) In 150 BC Jain mathematicians in India wrote the "Sthananga Sutra", which contains work on the theory of numbers, arithmetical operations, and operations with fractions.

A modern expression of fractions known as bhinnarasi seems to have originated in India in the work of Aryabhatta (c. AD 500), Brahmagupta (c. 628), and Bhaskara (c. 1150). Their works form fractions by placing the numerators (Sanskrit: amsa) over the denominators (cheda), but without a bar between them. In Sanskrit literature, fractions were always expressed as an addition to or subtraction from an integer. The integer was written on one line and the fraction in its two parts on the next line. If the fraction was marked by a small circle è'f é' or cross è'+é', it is subtracted from the integer; if no such sign appears, it is understood to be added. For example, Bhaskara I writes:

६ १ २

१ १ १॰

४ ५ ९

which is the equivalent of

6 1 2

1 1 −1

4 5 9

and would be written in modern notation as 61/4, 11/5, and 2 ″ 1/9 (i.e., 18/9).

The horizontal fraction bar is first attested in the work of Al-Hassâr (fl. 1200), a Muslim mathematician from Fez, Morocco, who specialized in Islamic inheritance jurisprudence. In his discussion he writes, "... for example, if you are told to write three-fifths and a third of a fifth, write thus, $\frac{3\ \ 1}{5\ \ 3}$. The same fractional notation—with the fraction given before the integer—appears soon after in the work of Leonardo Fibonacci in the 13th century.

In discussing the origins of decimal fractions, Dirk Jan Struik states:

"The introduction of decimal fractions as a common computational practice can be dated back to the Flemish pamphlet De Thiende, published at Leyden in 1585, together with a French translation, La Disme, by the Flemish mathematician Simon Stevin (1548–1620), then settled in the Northern Netherlands. It is true that decimal fractions were used by the Chinese many centuries before Stevin and that the Persian astronomer Al-Kâshî used both decimal and sexagesimal fractions with great ease in his Key to arithmetic (Samarkand, early fifteenth century)."

While the Persian mathematician Jamshîd al-Kâshî claimed to have discovered decimal fractions himself in the 15th century, J. Lennart Berggren notes that he was mistaken, as decimal fractions were first used five centuries before him by the Baghdadi mathematician Abu'l-Hasan al-Uqlidisi as early as the 10th century.[n 2]

In formal education

Pedagogical tools

In primary schools, fractions have been demonstrated through Cuisenaire rods, Fraction Bars, fraction strips, fraction circles, paper (for folding or cutting), pattern blocks, pie-shaped pieces, plastic rectangles, grid paper, dot paper, geoboards, counters and computer software.

Documents for teachers

Several states in the United States have adopted learning trajectories from the Common Core State Standards Initiative's guidelines for mathematics education. Aside from sequencing the learning of fractions and operations with fractions, the document provides the following definition of a fraction: "A number expressible in the form a/b where a is a whole number and b is a positive whole number. (The word fraction in these standards always refers to a non-negative number.)" The document itself also refers to negative fractions.

MIXED NUMBERS

A mixed numeral expresses the exact same information as an integer written next to a fraction that is less than one. The mixed numeral is the combination and fractional presentation of a whole number and valid number.

So, for example, 5S! is the mixed numeral equivalent of 16/3.

A mixed number is a whole number, and a proper fraction represented together. It generally represents a number between any two whole numbers.

Look at the given image, it represents a fraction that is greater than 1 but less than 2. It is thus, a mixed number.

and would be written in modern notation as 61/4, 11/5, and 2 " 1/9 (i.e., 18/9).

The horizontal fraction bar is first attested in the work of Al-Hassâr (fl. †1200), a Muslim mathematician from Fez, Morocco, who specialized in Islamic inheritance jurisprudence. In his discussion he writes, "... for example, if you are told to write three-fifths and a third of a fifth, write thus, $\frac{3\ \ 1}{5\ \ 3}$. The same fractional notation—with the fraction given before the integer—appears soon after in the work of Leonardo Fibonacci in the 13th century.

In discussing the origins of decimal fractions, Dirk Jan Struik states:

"The introduction of decimal fractions as a common computational practice can be dated back to the Flemish pamphlet De Thiende, published at Leyden in 1585, together with a French translation, La Disme, by the Flemish mathematician Simon Stevin (1548–1620), then settled in the Northern Netherlands. It is true that decimal fractions were used by the Chinese many centuries before Stevin and that the Persian astronomer Al-Kâshî used both decimal and sexagesimal fractions with great ease in his Key to arithmetic (Samarkand, early fifteenth century)."

While the Persian mathematician Jamshîd al-Kâshî claimed to have discovered decimal fractions himself in the 15th century, J. Lennart Berggren notes that he was mistaken, as decimal fractions were first used five centuries before him by the Baghdadi mathematician Abu'l-Hasan al-Uqlidisi as early as the 10th century.[n 2]

In formal education

Pedagogical tools

In primary schools, fractions have been demonstrated through Cuisenaire rods, Fraction Bars, fraction strips, fraction circles, paper (for folding or cutting), pattern blocks, pie-shaped pieces, plastic rectangles, grid paper, dot paper, geoboards, counters and computer software.

Documents for teachers

Several states in the United States have adopted learning trajectories from the Common Core State Standards Initiative's guidelines for mathematics education. Aside from sequencing the learning of fractions and operations with fractions, the document provides the following definition of a fraction: "A number expressible in the form a/b where a is a whole number and b is a positive whole number. (The word fraction in these standards always refers to a non-negative number.)" The document itself also refers to negative fractions.

MIXED NUMBERS

A mixed numeral expresses the exact same information as an integer written next to a fraction that is less than one. The mixed numeral is the combination and fractional presentation of a whole number and valid number.

So, for example, 5S‡ is the mixed numeral equivalent of 16/3.

A mixed number is a whole number, and a proper fraction represented together. It generally represents a number between any two whole numbers.

Look at the given image, it represents a fraction that is greater than 1 but less than 2. It is thus, a mixed number.

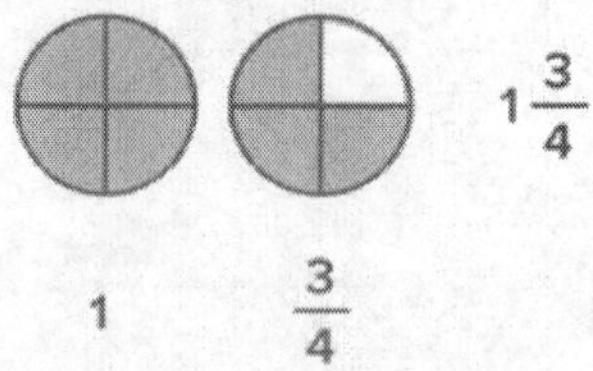

Some other examples of mixed numbers are

$3\frac{1}{2}$ $2\frac{3}{5}$ $5\frac{3}{5}$ $4\frac{1}{5}$

Parts of a Mixed Number

A mixed number is formed by combining three parts: a whole number, a numerator, and a denominator. The numerator and denominator are part of the proper fraction that makes the mixed number.

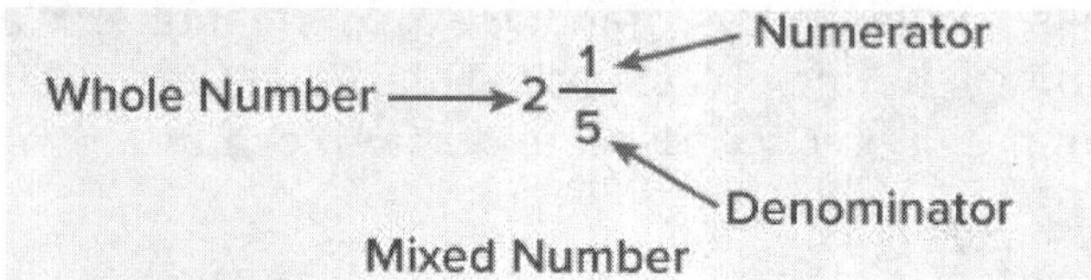

Properties of Mixed Numbers

- It is partly a whole number.
- It is partly a fraction.

Converting improper fractions to mixed fractions.

Step 1: Divide the numerator by the denominator.

Step 2: Write down the quotient as the whole number.

Step 3: Write down the remainder as the numerator and the divisor as the denominator.

For example, we follow the given steps to convert 7/3 into a mixed number form.

Step 1: Divide 7 by 3

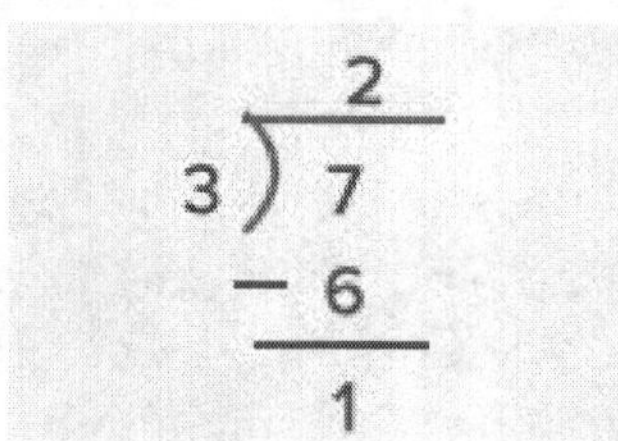

Step 2: Write quotient, divisor and remainder in form as in step 2 and step 3 above.

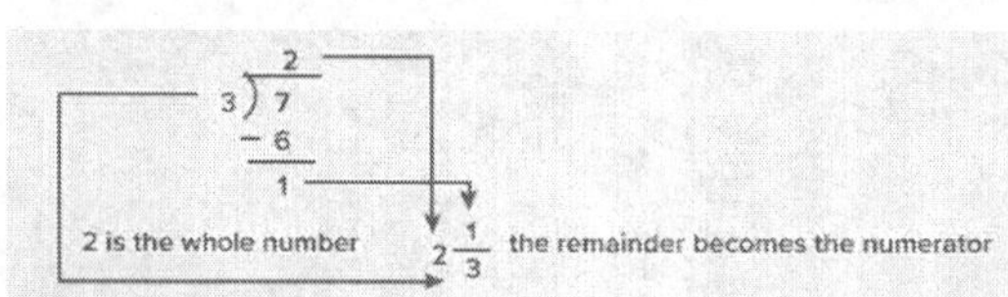

Adding Mixed Numbers

One can add mixed numbers by rearranging the whole numbers, adding them separately and adding the leftover fractions individually and in the end combing them all.

$1\ {}^{1}D_{2} + 3\ {}^{3}D_{4}$

Adding the whole numbers separately and the fractions separately.

For whole numbers:

1+3 = 4

For fractions: Find the LCM and then add

$$\frac{1\times1}{2\times2} + \frac{3}{4} = \frac{5}{4} = 1\frac{1}{4}$$

In the end, adding both the parts together.

$4+1\ {}^{1}D_{4} = 5\ {}^{1}D_{4}$

When a whole number and proper function combine and are represented in a new way, it is called a Mixed Number.

It is generally used to represent a number between any two whole numbers.

Example of Mixed Number

The diagram below represents a fraction that is greater than 1 but less than 2. Thus, it is a mixed number. That fraction which is more significant than one, but less than two is considered a mixed number.

The student will understand more about the mixed number when they know some examples. Some examples are 2 ½, 3 T! etc.

What is the Meaning of Mixed Fraction ?

In mathematics, a Mixed Number is the combination of a whole number and a fraction. The numerator and denominator are part of the proper fraction that makes the mixed number. D8

It is two-part in a fraction that helps to make the mixed number. Those are

- Denominator
- Numerator

So, a mixed number is partly a whole number and partly a fraction.

The Formula of Mixed Fraction

One must flow the given steps one after another to convert an improper fraction to a mixed fraction, which is called the mixed fraction formula. Consider the number 7/3.

- Step 1: Divide the numerator of the fraction with its denominator. That is 7/3.
- Step 2: The integer part that the student will get in the answer will be considered the integer part of the fraction. 2 is an integer in this case.
- Step 3: The denominator in the fraction will not change. It will be considered as the same given in the first. That is 3.
- Step 4: After applying all the steps correctly, an improper fraction that is 7/3 will change into a Mixed fraction that is 2S!

Pictorial representation of mixed number 2S!

How to Add Mixed Fractions?

We can either add the same denominators for both the fractions or the denominators can differ too. Here we have given a stepwise method to add the improper fraction with the same or different denominators. One can easily follow the given steps below to get the correct result.

Note: Before applying any arithmetic operations such as addition, subtraction, multiplication, etc., we need to change the mixed fractions to improper fractions.

Adding With the Same Denominators

Add: 6/4 + 5/4

- Step 1: Keep the denominator of the fractions the same as given the question. In that case which is 4.
- Step 2: Add the numerators of the fractions correctly.In that case which is 6+5=11.
- Step 3: You can get your answer in improper fractions. In that case, change it with Mixed Fraction. In that case 11/4 = 2(¾).

  So, We have 2 (¾) holes.

Adding With the Different Denominators

Add: 8/6 + 12/8

- Step 1: Calculate the LCM of the denominators of the given fraction. That is 24, LCM of 6 and 8.
- Step 2: In the next step, Multiply denominators of fractions and numerators of both fractions with a number such that they have the LCM as their new denominator. That is 8/6 by 4 and 12/8 by 3.
- Step 3: Add the new numerator properly and keep the denominator the same. That is 32 / 24 + 36 / 24 = 68/24 =17/6
- Step 4: You can get your answer in improper fractions. In that case, change it with Mixed Fraction. That is 2 (Z!).

  Subtracting Mixed Fractions:

Here's a step-wise explanation of how to Subtract the improper fraction with the Same or Different Denominators. Subtracting with the same Denominators.

Example: 6/4 – 5/4

Step 1: Keep the denominator '4' the same.

Step 2: Subtract the numerators '6' -'5' = 1.

Step 3: If the obtained answer is in the improper form then convert it into a mixed fraction. i.e. ¼

Subtracting With the Different Denominator

Subtract 12/8 – 8/6

Step 1: Find the LCM of the denominators, i.e. the LCM of 8 and 6 is 24

Step 2: Multiply denominators and numerators of both fractions with a number such that they have the LCM as their new denominator.Multiply the numerator and denominator of 8/6 by 4 and 12/8 by 3.

Step 3: Subtract the numerator and keep the denominators the same as it is.36 / 24 – 32/24 = 4/24

Step 4: If the obtained answer is in an improper form then convert it into Mixed Fraction. 4/24 = 1/6.

Multiplication of Mixed Fractions

Example: 2(Z!) × 3(½)

Step 1: Convert the given mixed fraction into an improper fraction. 17/ 6 × 7/2

Step 2: Multiply the numerators of both the fractions together and in similar way denominators of both the fractions together. (17 × 7)/(6 × 2)

Step 3: Now convert the fraction into the simplest form or Mixed fraction = 119/12 or 9(11/12).

Facts: Mixed number is also known as mixed fractions.

PRIME FACTORIZATION AND THE GCF

It is often useful to write a number in terms of its prime factorization, or as the product of its prime factors. For example, 56 can be written as 2×2×2×7 and 84 can be written as 2×2×3×7. Every number can be written as a product of primes, and, like a fingerprint, every number has a unique prime factorization.

To take a prime factorization of a number, start by dividing the number by its lowest prime factor. Write down this factor, and divide the new number by its lowest prime factor (it does not matter if this is the same as the first prime factor). Write this factor down and divide the new number by its lowest factor. Continue in this manner until the resulting number is prime. Write this number down as the final factor.

Every whole number greater than 1 has a prime factorization — that is, the list of prime numbers (including repeats) that equal that number when multiplied together. For example, here are the prime factorizations of 14, 20, and 300:

14 = 2 × 7

20 = 2 × 2 × 5

300 = 2 × 2 × 2 × 3 × 5

Factor trees are a useful tool for finding the prime factorization of a number. For example, to find the prime factorization 30, first find a factor pair — a pair of numbers that, when multiplied together, equals 30:

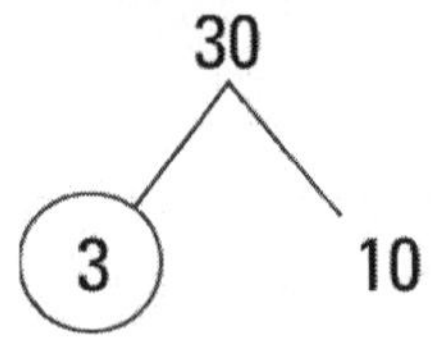

Note that 3 is circled because it's a prime number. But 10 is not prime, so you can continue the tree, finding a factor pair for 10:

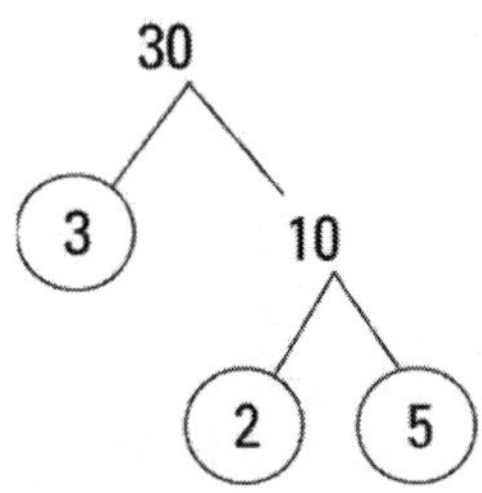

This time, both 2 and 5 are circled because they're each prime numbers. When the bottom numbers on the factor tree are all prime, you have your answer: $30 = 2 \times 3 \times 5$.

Using prime factorization to find the GCF

You can use prime factorization to find the greatest common factor (GCF) of a set of numbers. This method often works better for large numbers, when generating lists of all factors can be time-consuming.

Example 1: Compute the prime factorization of 1,575.

Step 1. Is 1,575 divisible by 2? No. By 3? Yes. $1,575/3 = 525$. Write down 3.

Step 2. Is 525 divisible by 3? Yes. $525/3 = 175$. Write down 3.

Step 3. Is 175 divisible by 3? No. By 5? Yes. $175/5 = 35$. Write down 5.

Step 4. Is 35 divisible by 5? Yes. $35/5 = 7$. Write down 5.

Step 5. 7 is prime. Write down 7.

Therefore, the prime factorization of 1,575 is $3 \times 3 \times 5 \times 5 \times 7$.

Example 2. Compute the prime factorization of 23,100.

Step 1. $23,100/2 = 11,550$. Write down 2.

Step 2. $11,550/2 = 5,775$. Write down 2.

Step 3. $5,775/3 = 1,925$. Write down 3.

Step 4. $1,925/5 = 385$. Write down 5.

Step 5. $385/5 = 77$. Write down 5.

Step 6. 77/7 = 11. Write down 7.

Step 7. 11 is prime. Write down 11.

Therefore, the prime factorization of 23,100 is $2 \times 2 \times 3 \times 5 \times 5 \times 7 \times 11$.

Greatest Common Factor

A common factor of two numbers is a factor that divides both numbers. The greatest common factor (GCF) of two numbers is the greatest number that divides both numbers. To find the GCF, take the prime factorization of both numbers. Then write down the factors that they have in common. If they share more than one of the same factor (two 2's, for example), write them both down. Then multiply the factors they have in common.

For example, the greatest common factor of 1,575 and 23,100 is $3 \times 5 \times 5 \times 7 = 525$. 1,575 and 23,100 are both divisible by 525, and they are not both divisible by any number greater than 525.

Relative Primes

Sometimes, two numbers do not have any prime factors in common. For example, the prime factorization of 40 is $2 \times 2 \times 2 \times 5$ and the prime factorization of 21 is 3×7. Since 40 and 21 have no common prime factors, they are said to be relatively prime, and their greatest common factor is 1.

6

Linear Equations

In mathematics, a linear equation is an equation that may be put in the form $a_1x_1 + \ldots + a_nx_n + b = 0$, where $x_1, \ldots, x_n$ are the variables (or unknowns), and $b, a_1, \ldots, a_n$ are the coefficients, which are often real numbers. The coefficients may be considered as parameters of the equation, and may be arbitrary expressions, provided they do not contain any of the variables. To yield a meaningful equation, the coefficients $a_1, \ldots, a_n$ are required to not all be zero.

Alternatively, a linear equation can be obtained by equating to zero a linear polynomial over some field, from which the coefficients are taken.

The solutions of such an equation are the values that, when substituted for the unknowns, make the equality true.

In the case of just one variable, there is exactly one solution (provided that a 1 '" 0). Often, the term linear equation refers implicitly to this particular case, in which the variable is sensibly called the unknown.

In the case of two variables, each solution may be interpreted as the Cartesian coordinates of a point of the Euclidean plane. The solutions of a linear equation form a line in the Euclidean plane, and, conversely, every line can be viewed as the set of all solutions of a linear equation in two variables. This is the origin of the term linear for describing this type of equations. More generally, the solutions of a linear equation in n variables form a hyperplane (a subspace of dimension n " 1) in the Euclidean space of dimension n.

Linear equations occur frequently in all mathematics and their applications in physics and engineering, partly because non-linear systems are often well approximated by linear equations.

This article considers the case of a single equation with coefficients from the field of real numbers, for which one studies the real solutions. All of its content applies to complex solutions and, more generally, for linear equations with coefficients and solutions in any field. For the case of several simultaneous linear equations, see system of linear equations.

A linear equation is an equation in which the highest power of the variable is always 1. It is also known as a one-degree equation. The standard form of a linear equation in one variable is of the form Ax + B = 0. Here, x is a variable, A is a coefficient and B is constant. The standard form of a linear equation in two variables is of the form Ax + By = C. Here, x and y are variables, A and B are coefficients and C is a constant.

An equation that has the highest degree of 1 is known as a linear equation. This means that no variable in a linear equation has an exponent more than 1. The graph of a linear equation always forms a straight line.

LINEAR EQUATION DEFINITION

A linear equation is an algebraic equation where each term has an exponent of 1 and when this equation is graphed, it always results in a straight line. This is the reason why it is named as a 'linear equation'.

There are linear equations in one variable and linear equations in two variables. Let us learn how to identify linear equations and non-linear equations with the help of the following examples.

Equations	Linear or Non-Linear
$y = 8x - 9$	Linear
$y = x^2 - 7$	Non-Linear, the power of the variable x is 2
$\sqrt{y} + x = 6$	Non-Linear, the power of the variable y is 1/2
$y + 3x - 1 = 0$	Linear
$y^2 - x = 9$	Non-Linear, the power of the variable y is 2

Linear Equation Formula

The linear equation formula is the way of expressing a linear equation. This can be done in different ways. For example, a linear equation can be expressed in the standard form, the slope-intercept form, or the point-slope form. Now, if we take the standard form of a linear equation, let us learn the way in which it is expressed. We can see that it varies from case to case based on the number of variables and it should be remembered that the highest (and the only) degree of all variables in the equation should be 1.

Linear Equations in Standard Form

The standard form or the general form of linear equations in one variable is written as, Ax + B = 0; where A and B are real numbers, and x is the single variable. The standard form of linear equations in two variables is expressed as, Ax + By = C; where A, B and C are any real numbers, and x and y are the variables.

Linear Equation Graph

The graph of a linear equation in one variable x forms a vertical line that is parallel to the y-axis and vice-versa, whereas, the graph of a linear equation in two variables x and y forms a straight line. Let us graph a linear equation in two variables with the help of the following example.

Example: Plot a graph for a linear equation in two variables, x - 2y = 2.

Let us plot the linear equation graph using the following steps.

- Step 1: The given linear equation is x - 2y = 2.
- Step 2: Convert the equation in the form of y = mx + b. This will give: y = x/2 - 1.
- Step 3: Now, we can replace the value of x for different numbers and get the resulting value of y to create the coordinates.
- Step 4: When we put x = 0 in the equation, we get y = 0/2 - 1, i.e. y = -1. Similarly, if we substitute the value of x as 2 in the equation, y = x/2 - 1, we get y = 0.
- Step 5: If we substitute the value of x as 4, we get y = 1. The value of x = -2 gives the value of y = -2. Now, these pairs of values of (x, y) satisfy the given linear equation y = x/2 - 1. Therefore, we list the coordinates as shown in the following table.

x	0	2	4	-2
y	-1	0	1	-2

- Step 6: Finally, we plot these points (4,1), (2,0), (0,-1) and (-2, -2) on a graph and join the points to get a straight line. This is how a linear equation is represented on a graph.

One variable

A linear equation in one variable x is of the form ax + b = 0, where a and b are real numbers and a ′ 0.

The root of $x = -\frac{b}{a}$.

Two variables

A linear equation in two variables x and y is of the form a x + by + c = 0, where a, b and c are real numbers such that $a^2 + b^2$ ′ 0.

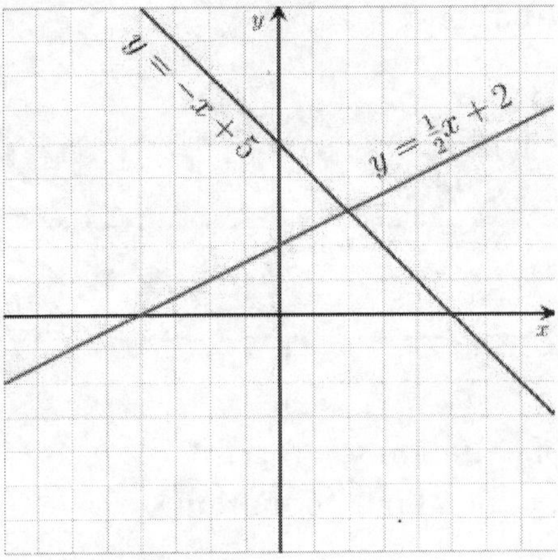

It has infinitely many possible solutions.

Linear function

If b ‘‴ 0, the equation

$ax + by + c = 0$

is a linear equation in the single variable y for every value of x. It has therefore a unique solution for y, which is given by

$$y = -\frac{a}{b}x - \frac{c}{b}.$$

This defines a function. The graph of this function is a line with slope $-\frac{a}{b}$ and y-intercept $-\frac{c}{b}$. The functions whose graph is a line are generally called linear functions in the context of calculus. However, in linear algebra, a linear function is a function that maps a sum to the sum of the images of the summands. So, for this definition, the above function is linear only when c = 0, that is when the line passes through the origin. For avoiding confusion, the functions whose graph is an arbitrary line are often called affine functions.

Geometric interpretation

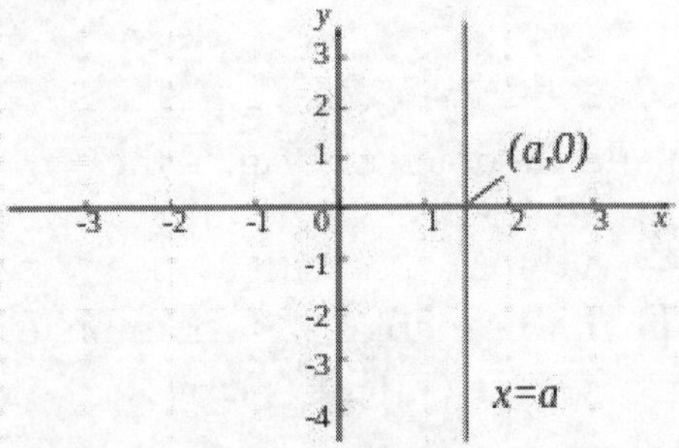

Vertical line of equation x = a

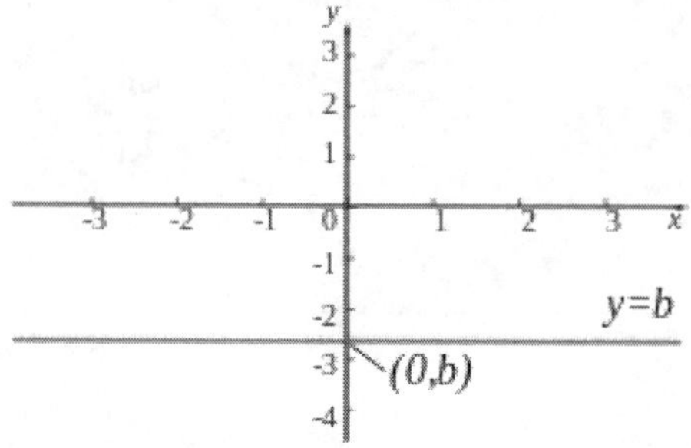

Horizontal line of equation y = b

Each solution (x, y) of a linear equation

ax + by + c = 0

may be viewed as the Cartesian coordinates of a point in the Euclidean plane. With this interpretation, all solutions of the equation form a line, provided that a and b are not both zero. Conversely, every line is the set of all solutions of a linear equation.

The phrase "linear equation" takes its origin in this correspondence between lines and equations: a linear equation in two variables is an equation whose solutions form a line.

If b '" 0, the line is the graph of the function of x that has been defined in the preceding section. If b = 0, the line is a vertical line (that is a line parallel to the y-axis) of equation $x = -\frac{c}{a}$, which is not the graph of a function of x.

Similarly, if a '" 0, the line is the graph of a function of y, and, if a = 0, one has a horizontal line of equation $y = -\frac{c}{b}$.

More than two variables

A linear equation with more than two variables may always be assumed to have the form

$$a_1x_1 + a_2x_2 + \cdots + a_nx_n + b = 0.$$

The coefficient b, often denoted a_0 is called the constant term (sometimes the absolute term in old books). Depending on the context, the term coefficient can be reserved for the a_i with i > 0.

When dealing with n = 3 variables, it is common to use x, y and z instead of indexed variables.

A solution of such an equation is a n-tuples such that substituting each element of the tuple for the corresponding variable transforms the equation into a true equality.

For an equation to be meaningful, the coefficient of at least one variable must be non-zero. In fact, if every variable has a zero coefficient, then, as mentioned for one variable, the equation is either inconsistent (for b '" 0) as having no solution, or all n-tuples are solutions.

The n-tuples that are solutions of a linear equation in n variables are the Cartesian coordinates of the points of an (n " 1)-dimensional hyperplane in an n-dimensional Euclidean space (or affine space if the coefficients are complex numbers or belong to any field). In the case of three variable, this hyperplane is a plane.

If a linear equation is given with a_j '" 0, then the equation can be solved for x_j, yielding

$$x_j = -\frac{b}{a_j} - \sum_{i\in\{1,\ldots,n\}, i\neq j} \frac{a_i}{a_j} x_i.$$

If the coefficients are real numbers, this defines a real-valued function of n real variables.

How to Solve Linear Equations?

An equation is like a weighing balance with equal weights on both sides. If we add or subtract the same number from both sides of an equation, it still holds true. Similarly, if we multiply or divide the same number on both sides of an equation, it is correct. We bring the variables to one side of the equation and the constant to the other side and then find the value of the unknown variable. This is the way to solve a linear equation with one variable. Let us understand this with the help of an example.

Example: Solve the equation, 3x - 2 = 4.

We perform mathematical operations on the Left-hand side (LHS) and the right-hand side (RHS) so that the balance is not disturbed. So, let us add 2 on both sides to reduce the LHS to 3x. This will not disturb the balance. The new LHS is 3x - 2 + 2 = 3x and the new RHS is 4 + 2 = 6. Now, let us divide both sides by 3 to reduce the LHS to x. Thus, we have x = 2. This is one of the ways of solving linear equations in one variable.

Tips on Linear Equations:

- The value of the variable that makes a linear equation true is called the solution or root of the linear equation.
- The solution of a linear equation is unaffected if the same number is added, subtracted, multiplied, or divided into both sides of the equation.

- The graph of a linear equation in one or two variables always forms a straight line.

Linear inequality

In mathematics a linear inequality is an inequality which involves a linear function. A linear inequality contains one of the symbols of inequality:. It shows the data which is not equal in graph form.

- < less than
- > greater than
- d″ less than or equal to
- e″ greater than or equal to
- '‴ not equal to
- = equal to

A linear inequality looks exactly like a linear equation, with the inequality sign replacing the equality sign.

Linear inequalities of real numbers

Two-dimensional linear inequalities

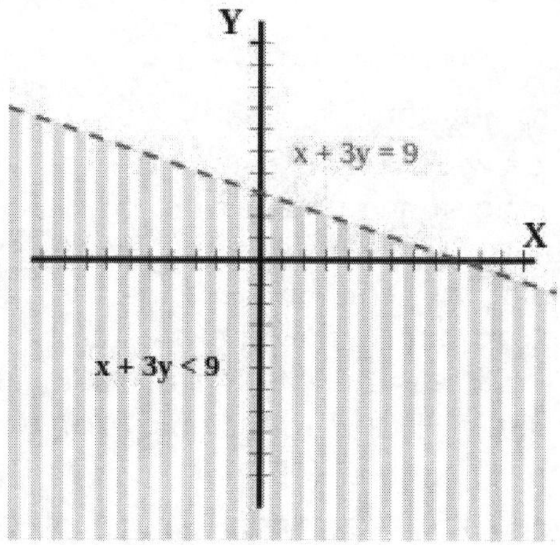

Graph of linear inequality: $x + 3y < 9$

Two-dimensional linear inequalities are expressions in two variables of the form:

ax + by < c and ax + by e″ c,

where the inequalities may either be strict or not. The solution set of such an inequality can be graphically represented by a half-plane (all the points on one "side" of a fixed line) in the Euclidean plane. The line that determines the half-planes ($ax + by = c$) is not included in the solution set when the inequality is strict. A simple procedure to determine which half-plane is in the solution set is to calculate the value of $ax + by$ at a point $(x_0,$

y_0) which is not on the line and observe whether or not the inequality is satisfied.

For example, to draw the solution set of x + 3y < 9, one first draws the line with equation x + 3y = 9 as a dotted line, to indicate that the line is not included in the solution set since the inequality is strict. Then, pick a convenient point not on the line, such as (0,0). Since 0 + 3(0) = 0 < 9, this point is in the solution set, so the half-plane containing this point (the half-plane "below" the line) is the solution set of this linear inequality.

Linear inequalities in general dimensions

In R^n linear inequalities are the expressions that may be written in the form

$$f(\overline{x}) < b \text{ or } f(\overline{x}) \leq b,$$

where f is a linear form (also called a linear functional), $\overline{x} = (x_1, x_2, \ldots, x_n)$ and b a constant real number.

More concretely, this may be written out as

$$a_1x_1 + a_2x_2 + \cdots + a_nx_n < b$$

or

$$a_1x_1 + a_2x_2 + \cdots + a_nx_n \leq b.$$

Here $x_1, x_2, ..., x_n$ are called the unknowns, and $a_1, a_2, ..., a_n$ are called the coefficients.

Alternatively, these may be written as

g (x) < 0 or g(x) d" 0,

where g is an affine function.

That is

$$a_0 + a_1x_1 + a_2x_2 + \cdots + a_nx_n < 0$$

or

$$a_0 + a_1x_1 + a_2x_2 + \cdots + a_nx_n \leq 0.$$

Note that any inequality containing a "greater than" or a "greater than or equal" sign can be rewritten with a "less than" or "less than or equal" sign, so there is no need to define linear inequalities using those signs.

Applications

Polyhedra

The set of solutions of a real linear inequality constitutes a half-space of the 'n'-dimensional real space, one of the two defined by the corresponding linear equation. The set of solutions of a system of linear inequalities corresponds to the intersection of the half-spaces defined by individual inequalities. It is a convex set, since the half-spaces are convex sets, and the intersection of a set of convex sets is also convex. In the non-degenerate cases this convex set is a convex polyhedron (possibly unbounded, e.g., a half-space, a slab between two parallel half-spaces or a polyhedral cone). It may also be empty or a convex polyhedron of lower dimension confined to an affine subspace of the n-dimensional space R^n.

Linear programming

A linear programming problem seeks to optimize (find a maximum or minimum value) a function (called the objective function) subject to a number of constraints on the variables which, in general, are linear inequalities. The list of constraints is a system of linear inequalities.

Generalization

The above definition requires well-defined operations of addition, multiplication and comparison; therefore, the notion of a linear inequality may be extended to ordered rings, and in particular to ordered fields.

7

Quadratic Equations

Quadratic equations are second-degree algebraic expressions and are of the form $ax^2 + bx + c = 0$. The word "Quadratic" is derived from the word "Quad" which means square. In other words, a quadratic equation is an "equation of degree 2." There are many scenarios where a quadratic equation is used. Did you know that when a rocket is launched, its path is described by a quadratic equation? Further, a quadratic equation has numerous applications in physics, engineering, astronomy,etc.

The quadratic equations are second-degree equations in x that have maximum two answers for x. These two answers for x are also called the roots of the quadratic equations and are designated as (á, â). We shall learn more about the roots of a quadratic equation in the below content.

A quadratic equation is an algebraic equation of the second degree in x. The quadratic equation in its standard form is $ax^2 + bx + c = 0$, where a and b are the coefficients, x is the variable, and c is the constant term. The first condition for an equation to be a quadratic equation is the coefficient of x^2 is a non-zero term(a '"0). For writing a quadratic equation in standard form, the x^2 term is written first, followed by the x term, and finally, the constant term is written. The numeric values of a, b, c are generally not written as fractions or decimals but are written as integral values.

In algebra, a quadratic equation (from Latin quadratus 'square') is any equation that can be rearranged in standard form as

$$ax^2 + bx + c = 0$$

where x represents an unknown, and a, b, and c represent known numbers, where a '" 0. If a = 0, then the equation is linear, not quadratic, as there is no ax^2 term. The numbers a, b, and c are the coefficients of the equation and may be distinguished by calling them, respectively, the quadratic coefficient, the linear coefficient and the constant or free term.

The values of x that satisfy the equation are called solutions of the equation, and roots or zeros of the expression on its left-hand side. A quadratic equation has at most two solutions. If there is only one solution, one says that it is a double root. If all the coefficients are real numbers, there are either two real solutions, or a single real double root, or two complex solutions. A quadratic equation always has two roots, if complex roots are included; and a double root is counted for two. A quadratic equation can be factored into an equivalent equation

$$ax^2 + bx + c = a(x - r)(x - s) = 0$$

where r and s are the solutions for x.

The quadratic formula

$$x = \frac{-b \pm \sqrt{b^2 - 4ac}}{2a}$$

expresses the solutions in terms of a, b, and c. Completing the square is one of several ways for getting it.

Solutions to problems that can be expressed in terms of quadratic equations were known as early as 2000 BC.

Because the quadratic equation involves only one unknown, it is called "univariate". The quadratic equation contains only powers of x that are non-negative integers, and therefore it is a polynomial equation. In particular, it is a second-degree polynomial equation, since the greatest power is two.

Quadratic Equation Formula

Quadratic Formula is the simplest way to find the roots of a quadratic equation. There are certain quadratic equations that cannot be easily factorized, and here we can conveniently use this quadratic formula to find the roots in the quickest possible way. The roots of the quadratic equation further help to find the sum of the roots and the product of the roots of the quadratic equation. The two roots in the quadratic formula are presented as a single expression. The positive sign and the negative sign can be alternatively used to obtain the two distinct roots of the equation.

The roots of a quadratic equation $ax^2 + bx + c = 0$ are given by x = [-b ± "(b^2 - 4ac)]/2a.

Important Formulas for Solving Quadratic Equations

The following list of important formulas is helpful to solve quadratic equations.

- The quadratic equation in its standard form is $ax^2 + bx + c = 0$
- The discriminant of the quadratic equation is $D = b^2 - 4ac$

- For D > 0 the roots are real and distinct.
- For D = 0 the roots are real and equal.
- For D < 0 the roots do not exist, or the roots are imaginary.
- The formula to find the roots of the quadratic equation is x = [-b ± "(b^2 - 4ac)]/2a.
- The sum of the roots of a quadratic equation is á + â = -b/a = - Coefficient of x/ Coefficient of x^2.
- The product of the Root of the quadratic equation is áâ = c/a = Constant term/ Coefficient of x^2
- The quadratic equation having roots á, â, is x^2 - (á + â)x + áâ = 0.
- The condition for the quadratic equations $a_1x^2 + b_1x + c_1 = 0$, and $a_2x^2 + b_2x + c_2 = 0$ having the same roots is $(a_1b_2 - a_2b_1)(b_1c_2 - b_2c_1) = (a_2c_1 - a_1c_2)^2$.
- For positive values of a (a > 0), the quadratic expression $f(x) = ax^2 + bx + c$ has a minimum value at x = -b/2a.
- For negative value of a (a < 0), the quadratic expression $f(x) = ax^2 + bx + c$ has a maximum value at x = -b/2a.
- For a > 0, the range of the quadratic equation $ax^2 + bx + c = 0$ is [b^2 - 4ac/ 4a, ")
- For a < 0, the range of the quadratic equation $ax^2 + bx + c = 0$ is : (", -(b^2 - 4ac)/4a]

Quadratic Formula Proof

Consider an arbitrary quadratic equation: $ax^2 + bx + c = 0$, a '" 0

To determine the roots of this equation, we proceed as follows:

$ax^2 + bx = -c$ ⇒! $x^2 + bx/a = -c/a$

Now, we express the left hand side as a perfect square, by introducing a new term $(b/2a)^2$ on both sides:

$x^2 + bx/a + (b/2a)^2 = -c/a + (b/2a)^2$

The left hand side is now a perfect square:

$(x + b/2a)^2 = -c/a + b^2/4a^2$ ⇒! $(x + b/2a)^2 = (b^2 - 4ac)/4a^2$

This is good for us, because now we can take square roots to obtain:

x + b/2a = ±"(b^2 - 4ac)/2a

x = (-b ± "(b^2 - 4ac))/2a

Thus, by completing the squares, we were able to isolate x and obtain the two roots of the equation.

Relationship Between Coefficients and Roots of Quadratic Equation

The coefficient of x^2, x term, and the constant term of the quadratic equation $ax^2 + bx + c = 0$ are useful to study more about the properties of roots of the quadratic equation. The sum and product of roots of a quadratic equation can be directly calculated from the equation, without actually finding the roots of the quadratic equation. The sum of the roots of the quadratic equation is equal to the negative of the coefficient of x divided by the coefficient of x^2. The product of the root of the equation is equal to the constant term divided by the coefficient of the x^2. For a quadratic equation $ax^2 + bx + c = 0$, the sum and product of the roots are as follows.

- Sum of the Roots: á + â = -b/a = - Coefficient of x/ Coefficient of x^2
- Product of the Roots: áâ = c/a = Constant term/ Coefficient of x^2

The quadratic equation can also be formed for the given roots of the equation. If á, â, are the roots of the quadratic equation, then the quadratic equation is as follows.

x^2 - (á + â)x + áâ = 0

Methods to Solve Quadratic Equations

A quadratic equation can be solved to obtain two values of x or the two roots of the equation. There are four different methods to find the roots of the quadratic equation. The four methods of solving the quadratic equations are as follows.

- Factorizing of Quadratic Equation
- Formula Method of Finding Roots
- Method of Completing the Square
- Graphing Method to Find the Roots

Let us look in detail at each of the above methods to understand how to use these methods, their applications, and their uses.

Factorization of Quadratic Equation

Factorization of quadratic equation follows a sequence of steps. For a general form of the quadratic equation $ax^2 + bx + c = 0$, we need to first split the middle term into two terms, such that the product of the terms is equal to the constant term. Further, we can take the common terms from the available term, to finally obtain the required factors. For understanding factorization, the general form of the quadratic equation can be presented as follows.

- $x^2 + (a + b)x + ab = 0$
- $x^2 + ax + bx + ab = 0$
- $x(x + a) + b(x + a)$
- $(x + a)(x + b) = 0$

Let us understand factorization through the below example.

- $x^2 + 5x + 6 = 0$
- $x^2 + 2x + 3x + 6 = 0$
- $x(x + 2) + 3(x + 2) = 0$

- $(x + 2)(x + 3) = 0$

Thus the two obtained factors of the quadratic equation are $(x + 2)$ and $(x + 3)$. To find its roots, just set each factor to zero and solve for x. i.e., $x + 2 = 0$ and $x + 3 = 0$ which gives $x = -2$ and $x = -3$. Thus, $x = -2$ and $x = -3$ are the roots of $x^2 + 5x + 6 = 0$.

Quadratic Formula to Find Roots

The quadratic equations which cannot be solved through the method of factorization can be solved with the help of a formula. The formula to solve the quadratic equation uses the terms from the standard form of a quadratic equation. Through the below formula we can obtain the two roots of x by first using the positive sign in the formula and then using the negative sign. Any quadratic equation can be solved using this formula.

Method of Completing the Square

The method of completing the square for a quadratic equation, is to algebraically square and simplify, to obtain the required roots of the equation. Consider a quadratic equation $ax^2 + bx + c = 0$, a '" 0. To determine the roots of this equation, we simplify it as follows:

- $ax^2 + bx + c = 0$
- $ax^2 + bx = -c$
- $x^2 + bx/a = -c/a$

Now, we express the left hand side as a perfect square, by introducing a new term $(b/2a)^2$ on both sides:

- $x^2 + bx/a + (b/2a)^2 = -c/a + (b/2a)^2$
- $(x + b/2a)^2 = -c/a + b^2/4a^2$
- $(x + b/2a)^2 = (b^2 - 4ac)/4a^2$
- x + b/2a = +"(b^2- 4ac)/2a
- x = - b/2a +"(b^2- 4ac)/2a

Now with this method of completing the square, we could consolidate the value for the roots of the equation. Further on simplification and taking the square root, the two possible roots of the quadratic equation are, x = (-b + "(b^2- 4ac))/2a. Here the '+' sign gives one root and the '-' sign gives another root of the quadratic equation. Generally, this detailed method is avoided, and only the quadratic formula is used to obtain the required roots.

Quadratic Equations Having Common Roots

Let the two quadratic equations having common roots are $a_1x^2 + b_1x + c_1 = 0$, and $a_2x^2 + b_2x + c_2 = 0$. Let us solve these two equations to find the conditions for which these equations have a common root. The two equations are solved for x^2 and x respectively.

$(x^2)(b_1c_2 - b_2c_1) = (-x)/(a_1c_2 - a_2c_1) = 1/(a_1b_2 - a_2b_1)$

$x^2 = (b_1c_2 - b_2c_1) / (a_1b_2 - a_2b_1)$

$x = (a_2c_1 - a_1c_2) / (a_1b_2 - a_2b_1)$

Hence, by simplifying the above two expressions we have the following condition for the two equations having the common root.

$(a_1b_2 - a_2b_1)(b_1c_2 - b_2c_1) = (a_2c_1 - a_1c_2)^2$

Maximum and Minimum Value of Quadratic Expression

The maximum and minimum value for the quad equation $F(x) = ax^2 + bx + c = 0$ can be observed in the below graphs. For positive values of a ($a > 0$), the quadratic expression has a minimum value at $x = -b/2a$, and for negative value of a ($a < 0$), the quadratic expression has a maximum value at $x = -b/2a$.

The maximum and minimum values of the quadratic expressions are of further help to find the range of the quadratic expression: The range of the quadratic expressions also depends on the value of a. For positive values of a(a > 0), the range is [F(-b/2a), "), and for negative values of a (a < 0), the range is (-", F(-b/2a)].

- For a > 0, Range: [f(-b/2a), ")
- For a < 0, Range: (-", f(-b/2a)]

Note that the domain of a quadratic function is the set of all real numbers, i.e., (-", ").

Roots of a Quadratic Equation

The roots of a quadratic equation are the two values of x, which are obtained by solving the quadratic equation. The roots of a quadratic equation are referred to by the symbols alpha (á), and beta (â). These roots of the quadratic equation are also called the zeros of the equation. Here we shall learn more about how to find the nature of roots of a quadratic equation without actually finding the roots of the equation. And also check out the formulas to find the sum and the product of the roots of the equation.

Nature of Roots of the Quadratic Equation

The nature of roots of a quadratic equation can be found without actually finding the roots (á, â) of the equation. This is possible by taking the discriminant value, which is part of the formula to solve the quadratic equation. The value $b^2 - 4ac$ is called the discriminant of a quadratic equation and is designated as 'D'. Based on the discriminant value the nature of the roots of the quadratic equation can be predicted.

Discriminant: $D = b^2 - 4ac$

- D > 0, the roots are real and distinct

- $D = 0$, the roots are real and equal.
- $D < 0$, the roots do not exist or the roots are imaginary.

Solving the quadratic equation

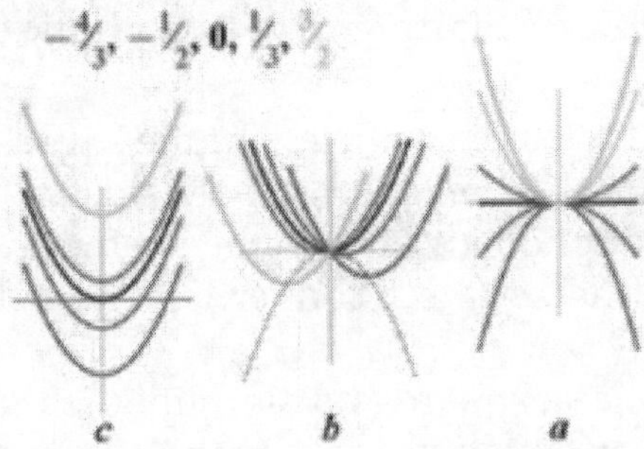

Plots of quadratic function $y = ax^2 + bx + c$, varying each coefficient separately while the other coefficients are fixed (at values $a = 1, b = 0, c = 0$)

A quadratic equation with real or complex coefficients has two solutions, called roots. These two solutions may or may not be distinct, and they may or may not be real.

Factoring by inspection

It may be possible to express a quadratic equation $ax^2 + bx + c = 0$ as a product $(px + q)(rx + s) = 0$. In some cases, it is possible, by simple inspection, to determine values of p, q, r, and s that make the two forms equivalent to one another. If the quadratic equation is written in the second form, then the "Zero Factor Property" states that the quadratic equation is satisfied if $px + q = 0$ or $rx + s = 0$. Solving these two linear equations provides the roots of the quadratic.

For most students, factoring by inspection is the first method of solving quadratic equations to which they are exposed.[: 202–207] If one is given a quadratic equation in the form $x^2 + bx + c = 0$, the sought factorization has the form $(x + q)(x + s)$, and one has to find two numbers q and s that add up to b and whose product is c (this is sometimes called "Vieta's rule" and is related to Vieta's formulas). As an example, $x^2 + 5x + 6$ factors as $(x + 3)(x + 2)$. The more general case where a does not equal 1 can require a considerable effort in trial and error guess-and-check, assuming that it can be factored at all by inspection.

Except for special cases such as where $b = 0$ or $c = 0$, factoring by inspection only works for quadratic equations that have rational roots. This means that the great majority of quadratic equations that arise in practical applications cannot be solved by factoring by inspection.[: 207]

Completing the square

The process of completing the square makes use of the algebraic identity

$$x^2 + 2hx + h^2 = (x + h)^2,$$

which represents a well-defined algorithm that can be used to solve any quadratic equation.: 207 Starting with a quadratic equation in standard form, $ax^2 + bx + c = 0$

1. Divide each side by a, the coefficient of the squared term.
2. Subtract the constant term c/a from both sides.
3. Add the square of one-half of b/a, the coefficient of x, to both sides. This "completes the square", converting the left side into a perfect square.
4. Write the left side as a square and simplify the right side if necessary.
5. Produce two linear equations by equating the square root of the left side with the positive and negative square roots of the right side.
6. Solve each of the two linear equations.

We illustrate use of this algorithm by solving $2x^2 + 4x$ " $4 = 0$

1) $x^2 + 2x - 2 = 0$

2) $x^2 + 2x = 2$

3) $x^2 + 2x + 1 = 2 + 1$

4) $(x+1)^2 = 3$

5) $x + 1 = \pm\sqrt{3}$

6) $x = -1 \pm \sqrt{3}$

The plus–minus symbol "±" indicates that both x = "1 + "3 and x = "1 " "3 are solutions of the quadratic equation.

Quadratic formula and its derivation

Completing the square can be used to derive a general formula for solving quadratic equations, called the quadratic formula. The mathematical proof will now be briefly summarized. It can easily be seen, by polynomial expansion, that the following equation is equivalent to the quadratic equation:

$$\left(x + \frac{b}{2a}\right)^2 = \frac{b^2 - 4ac}{4a^2}.$$

Taking the square root of both sides, and isolating x, gives:

$$x = \frac{-b \pm \sqrt{b^2 - 4ac}}{2a}.$$

Some sources, particularly older ones, use alternative parameterizations of the quadratic equation such as $ax^2 + 2bx + c = 0$ or ax^2 " $2bx + c = 0$, where b has a magnitude one half of the more common one, possibly with opposite sign. These result in slightly different forms for the solution, but are otherwise equivalent.

A number of alternative derivations can be found in the literature. These proofs are simpler than the standard completing the square method, represent interesting applications of other frequently used techniques in algebra, or offer insight into other areas of mathematics.

A lesser known quadratic formula, as used in Muller's method provides the same roots via the equation

$$x = \frac{2c}{-b \pm \sqrt{b^2 - 4ac}}.$$

This can be deduced from the standard quadratic formula by Vieta's formulas, which assert that the product of the roots is c/a.

One property of this form is that it yields one valid root when a = 0, while the other root contains division by zero, because when a = 0, the quadratic equation becomes a linear equation, which has one root. By contrast, in this case, the more common formula has a division by zero for one root and an indeterminate form 0/0 for the other root. On the other hand, when c = 0, the more common formula yields two correct roots whereas this form yields the zero root and an indeterminate form 0/0.

Reduced quadratic equation

It is sometimes convenient to reduce a quadratic equation so that its leading coefficient is one. This is done by dividing both sides by a, which is always possible since a is non-zero. This produces the reduced quadratic equation:

$$x^2 + px + q = 0,$$

where p = b/a and q = c/a. This monic polynomial equation has the same solutions as the original.

The quadratic formula for the solutions of the reduced quadratic equation, written in terms of its coefficients, is:

$$x = \frac{1}{2}\left(-p \pm \sqrt{p^2 - 4q}\right),$$

or equivalently:

$$x = -\frac{p}{2} \pm \sqrt{\left(\frac{p}{2}\right)^2 - q}.$$

Avoiding loss of significance

Although the quadratic formula provides an exact solution, the result is not exact if real numbers are approximated during the computation, as usual in numerical analysis, where real numbers are approximated by floating point numbers (called "reals" in many programming languages). In this context, the quadratic formula is not completely stable.

This occurs when the roots have different order of magnitude, or, equivalently, when b^2 and b^2 “ 4ac are close in magnitude. In this case, the subtraction of two nearly equal numbers will cause loss of significance or catastrophic cancellation in the smaller root. To avoid this, the root that is smaller in magnitude, r, can be computed as (c/a)/R where R is the root that is bigger in magnitude.

A second form of cancellation can occur between the terms b^2 and 4ac of the discriminant, that is when the two roots are very close. This can lead to loss of up to half of correct significant figures in the roots.

History

Babylonian mathematicians, as early as 2000 BC (displayed on Old Babylonian clay tablets) could solve problems relating the areas and sides of rectangles. There is evidence dating this algorithm as far back as the Third Dynasty of Ur. In modern notation, the problems typically involved solving a pair of simultaneous equations of the form:

$x + y = p, \quad xy = q,$

which is equivalent to the statement that x and y are the roots of the equation:[:86]

$z^2 + q = pz.$

The steps given by Babylonian scribes for solving the above rectangle problem, in terms of x and y, were as follows:

1. Compute half of p.
2. Square the result.
3. Subtract q.
4. Find the (positive) square root using a table of squares.
5. Add together the results of steps (1) and (4) to give x.

In modern notation this means calculating $x = \left(\frac{p}{2}\right) + \sqrt{\left(\frac{p}{2}\right)^2 - q}$, which is equivalent to the modern day quadratic formula for the larger real root (if any) $x = \frac{-b + \sqrt{b^2 - 4ac}}{2a}$ with a = 1, b = "p, and c = q.

Geometric methods were used to solve quadratic equations in Babylonia, Egypt, Greece, China, and India. The Egyptian Berlin Papyrus, dating back to the Middle Kingdom (2050 BC to 1650 BC), contains the solution to a two-term quadratic equation. Babylonian mathematicians from circa 400 BC and Chinese mathematicians from circa 200 BC used geometric methods of dissection to solve quadratic equations with positive roots. Rules for quadratic equations were given in The Nine Chapters on the Mathematical Art, a Chinese treatise on mathematics. These early geometric methods do not appear to have had a general formula. Euclid, the Greek mathematician, produced a more abstract geometrical method around 300 BC. With a purely geometric approach Pythagoras and Euclid created a general procedure to find solutions of the quadratic equation. In his work Arithmetica, the Greek mathematician Diophantus solved the quadratic equation, but giving only one root, even when both roots were positive.

In 628 AD, Brahmagupta, an Indian mathematician, gave the first explicit (although still not completely general) solution of the quadratic equation $ax^2 + bx = c$ as follows: "To the absolute number multiplied by four times the [coefficient of the] square, add the square of the [coefficient of the] middle term; the square root of the same, less the [coefficient of the] middle term, being divided by twice the [coefficient of the] square is the value." (Brahmasphutasiddhanta, Colebrook translation, 1817, page 346):[87] This is equivalent to

$$x = \frac{\sqrt{4ac + b^2} - b}{2a}.$$

The Bakhshali Manuscript written in India in the 7th century AD contained an algebraic formula for solving quadratic equations, as well as quadratic indeterminate equations (originally of type ax/c = y). Muhammad ibn Musa al-Khwarizmi (9th century), possibly inspired by Brahmagupta, developed a set of formulas that worked for positive solutions. Al-Khwarizmi goes further in providing a full solution to the general quadratic equation, accepting one or two numerical answers for every quadratic equation, while providing geometric proofs in the process. He also described the method of completing the square and recognized that the discriminant must be positive,[:230] which was proven by his contemporary 'Abd al-Hamîd ibn Turk (Central Asia, 9th century) who gave geometric figures to prove that if the discriminant is negative, a quadratic equation has no solution.[:234]

While al-Khwarizmi himself did not accept negative solutions, later Islamic mathematicians that succeeded him accepted negative solutions,[:191] as well as irrational numbers as solutions. Abû Kâmil Shujâ ibn Aslam (Egypt, 10th century) in particular was the first to accept irrational numbers (often in the form of a square root, cube root or fourth root) as solutions to quadratic equations or as coefficients in an equation. The 9th century Indian mathematician Sridhara wrote down rules for solving quadratic equations.

The Jewish mathematician Abraham bar Hiyya Ha-Nasi (12th century, Spain) authored the first European book to include the full solution to the general quadratic equation. His solution was largely based on Al-Khwarizmi's work. The writing of the Chinese mathematician Yang Hui (1238–1298 AD) is the first known one in which quadratic equations with negative coefficients of 'x' appear, although he attributes this to the earlier Liu Yi. By 1545 Gerolamo Cardano compiled the works related to the quadratic equations. The quadratic formula covering all cases was first obtained by Simon Stevin in 1594. In 1637 René Descartes published La Géométrie containing the quadratic formula in the form we know today.

Examples and applications

The golden ratio is found as the positive solution of the quadratic equation x^2 " x " 1 = 0.

The equations of the circle and the other conic sections—ellipses, parabolas, and hyperbolas—are quadratic equations in two variables.

Given the cosine or sine of an angle, finding the cosine or sine of the angle that is half as large involves solving a quadratic equation.

The process of simplifying expressions involving the square root of an expression involving the square root of another expression involves finding the two solutions of a quadratic equation.

Descartes' theorem states that for every four kissing (mutually tangent) circles, their radii satisfy a particular quadratic equation.

The equation given by Fuss' theorem, giving the relation among the radius of a bicentric quadrilateral's inscribed circle, the radius of its circumscribed circle, and the distance between the centers of those circles, can be expressed as a quadratic equation for which the distance between the two circles' centers in terms of their radii is one of the solutions. The other solution of the same equation in terms of the relevant radii gives the distance between the circumscribed circle's center and the center of the excircle of an ex-tangential quadrilateral.

Critical points of a cubic function and inflection points of a quartic function are found by solving a quadratic equation.

Solving a Quadratic Equation - Tips and Tricks

Some of the below-given tips and tricks on quadratic equations are helpful to more easily solve quadratic equations.

1. The quadratic equations are generally solved through factorization. But in instances when it cannot be solved by factorization, the formula is used.
2. The roots of a quadratic equation are also called the zeroes of the equation.
3. For quadratic equations having negative discriminant values, the roots are represented with the help of complex numbers.
4. The sum and product of roots of a quadratic equation can be used to find higher algebraic expressions involving these roots.

How to Determine the Nature of the Roots of a Quadratic Equation?

Example 1: Find the values of k for which the quadratic expression $(x - k)(x - 10) + 1 = 0$ has integral roots.

Solution:

The given equation can be rewritten as, $x^2 - (10 + k)x + 1 + 10k = 0$.

$D = b^2 - 4ac = 100 + k^2 + 20k - 40k = k^2 - 20k + 96 = (k - 10)^2 - 4$

The quadratic equation will have integral roots, if the value of discriminant > 0, D is a perfect square, a = 1 and b and c are integers.

i.e. $(k - 10)^2 - D = 4$

Since discriminant is a perfect square. Hence, the difference of two perfect square in R.H.S will be 4 only when D = 0 and $(k - 10)^2 = 4$.

⇒! $k - 10 = \pm 2$. Therefore, the values of k = 8 and 12.

Example 2: Find the values of k such that the equation $p/(x + r) + q/(x - r) = k/2x$ has two equal roots.

Solution:

The given quadratic equation can be rewritten as:

$[2p + 2q - k]x^2 - 2r[p - q]x + r^2k = 0$

For equal roots, the discriminant (D) = 0, i.e. $b^2 - 4ac = 0$

Here, $a = [\ 2p + 2q - k\ ]$, $b = -2r\ [\ p - q\ ]$ and $c = r^2k$

$[-2r\ (p - q)]^2 - 4[(2p + 2q - k)\ (r^2k)] = 0$

$r^2(p - q)^2 - r^2k(2p + 2q - k) = 0$

Since r '" 0, therefore, $(p - q)^2 - k(2p + 2q - k) = 0$

$k^2 - 2(p + q)k + (p - q)^2$

k = 2(p+q) ± "[4(p + q)^2 – 4(p – q)]^2/2 = -(p + q) ± "4pq

4" The values of k = (p + q) ± 2"pq = ("p ± "q)^2

Example 3: Find the quadratic equation with rational coefficients when one root is 1/(2 + "5).

Solution:

If the coefficients are rational, then the irrational roots occur in conjugate pairs. Therefore, if one root is á = 1/(2 + "5) = "5 – 2, then the other root will be â = 1/(2 – "5) = -"5 – 2.

Sum of the roots á + â = -4 and product of roots á â = -1.

Thus, the required equation is $x^2 + 4x - 1 = 0$.

Example 4: Form a quadratic equation with real coefficients when one of its root is (3 – 2i).

Solution:

Since the complex roots always occur in pairs, so the other root is 3 + 2i. Therefore, by obtaining the sum and the product of the roots, we can form the required quadratic equation.

The sum of the roots is

$(3 + 2i) + (3 - 2i) = 6$. The product of the root is $(3 + 2i) \times (3 - 2i) = 9 - 4i^2 = 9 + 4 = 13$.

Hence, the equation is $x^2 - Sx + P = 0$

Therefore, $x^2 - 6x + 13 = 0$ is the required quadratic equation.

1. Example 1: Meghan is a fitness enthusiast and goes for a jog every morning. The park where she jogs is rectangular in shape and measures 12m × 8m. An environmentalist group plans to revamp the park and decides to build a pathway surrounding the park. This would increase the total area to 140 sq m. What will be the width of the pathway?

Solution:

Let's denote the width of the pathway by x.

Then, the length and breadth of the outer rectangle is (12+2x) m and (8+2x) m.

According to the question,

$(12 + 2x)(8 + 2x) = 140$

$2(6 + x)\,2(4 + x) = 140$

$(6 + x)(4 + x) = 35$

$24 + 6x + 4x + x^2 = 35$

$x^2 + 10x - 11 = 0$

$x^2 + 11x - x - 11 = 0$

$x(x + 11) - 1(x + 11) = 0$

$(x + 11)(x - 1) = 0$

$(x + 11) = 0$ and $(x - 1) = 0$

$x = -11$ and $x = 1$

Since length can't be negative, we take $x = 1$.

Answer: Therefore the width of the pathway is 1 m.

2. Example 2: Let's learn how a quadratic equation question finds its application in the field of motion. Rita throws a ball upwards from a platform that is 20m above the ground. The height of the ball from the ground at a time 't', is denoted by 'h'. Suppose $h = -4t^2 + 16t + 20$. Find the maximum height attained by the ball.

Solution:

We can rearrange the terms of the quadratic equation

$h = -4t^2 + 16t + 20$

in such a way that it is easy to find the maximum value of this equation.

$h = -4t^2 + 16t + 20$

$= -4t^2 + 16t + 20$

$= -4(t^2 - 4t - 5)$

$= -4((t - 2)^2 - 9)$

$= -4(t - 2)^2 + 144$

We should keep the value of $(t-2)^2$ minimum in order to find the maximum value of h.

So, the minimum value $(t-2)^2$ can take is 0.

Answer: Therefore the maximum height attained is 144m.

3. Example 3: Find the quadratic equation having the roots 5 and 8 respectively.

Solution:

The quadratic equation having the roots á, â, is $x^2 - (á + â)x + áâ = 0$.

Given á = 5, and â = 8.

Therefore the quadratic equation is:

$x^2 - (5 + 8)x + 5 \times 8 = 0$

$x^2 - 13x + 40 = 0$

Answer: Hence the required quaratic equation is $x^2 - 13x + 40 = 0$

4. Example 4: The quad equation $2x^2 + 9x + 7 = 0$ has roots á, â. Find the quadratic equation having the roots 1/á, and 1/â.

Solution:

The quadratic equation having roots that are reciprocal to the roots of the equation $ax^2 + bx + c = 0$, is $cx^2 + bx + a = 0$.

The given quadratic equation is 2x^{2} + 9x + 7 = 0.

Hence the required equation having reciprocal roots is 7x^{2} + 9x + 2 = 0.

Answer: Therefore the equation is 7x^{2} + 9x + 2 = 0.

Relationships between Coefficient and Roots of Quadratic Equation

If á and â are roots of a Quadratic Equation ax^{2} + bx + c = 0 then,

- á + â = -b/a
- áâ = c/a
- á – â = ±"[(á + â)^{2} – 4áâ]
- |á + â| = "D/|a|

The relationship between the roots and coefficient of a polynomial equation can be derived by simplifying the given polynomials and substituting the above results as shown below.

- á^{2}â + â^{2}á = áâ (á + â) = – bc/a^{2}
- á^{2} + áâ + â^{2} = (á + â)^{2} – áâ = (b^{2} – ac)/a^{2}
- á^{2} + â^{2} = (á – â)^{2} – 2áâ
- á^{2} – â^{2} = (á + â) (á – â)
- á^{3} + â^{3} = (á + â)^{3} – 3áâ(á + â)
- á^{3} – â^{3} = (á – â)^{3} + 3áâ(á – â)
- (á/â)^{2} + (â/á)^{2} = á^{4} + â^{4}/á^{2}â^{2}

Example: If the coefficient of x in the quadratic equation x^{2} + bx + c =0 was taken as 17 in place of 13, its roots were found to be -2 and -15. Find the roots of the original quadratic equation.

Solution:

Since there is no change in the coefficient of x^{2} and c, therefore, the product of zeroes will remain the same for both equations.

Therefore, the product of zeroes (c) = -2 × -15 = 30,

Since, the original value of b is 13.

4" Sum of zeroes = -b/a = -13.

Hence, the original quadratic equation is:

x^{2} – (Sum of Zeroes)x + (Product of Zeroes) = 0

x^{2} + 13x + 30 = 0

4" (x + 10) (x + 3) = 0

Therefore, the roots of the original quadratic equations are -3 and -10.

8

Surds: Definition and Types

Numbers that are irrational and cannot be represented in the form of fractions or as recurring decimal numbers are known as surds. These numbers cannot be represented as recurring decimals or as fractions, they can be only represented as square roots. In other words, surds are nothing but square root representations of irrational numbers that cannot be expressed in fractional or recurring decimals. In order to make precise calculations, surds are used. In this article, you will be learning about the definition of surds, the basic six rules of the surds, some solved example problems, and the frequently asked questions related to surds.

In Mathematics, surds are the values in square root that cannot be further simplified into whole numbers or integers. Surds are irrational numbers. The examples of surds are "2, "3, "5, etc., as these values cannot be further simplified. If we further simply them, we get decimal values, such as:

"2 = 1.4142135…

"3 = 1.7320508…

"5 = 2.2360679…

In mathematics, an nth root of a number x is a number r which, when raised to the power n, yields x:

$r^n = x,$

where n is a positive integer, sometimes called the degree of the root. A root of degree 2 is called a square root and a root of degree 3, a cube root. Roots of higher degree are referred by using ordinal numbers, as in fourth root, twentieth root, etc. The computation of an nth root is a root extraction.

For example, 3 is a square root of 9, since $3^2 = 9$, and "3 is also a square root of 9, since $("3)^2 = 9$.

Any non-zero number considered as a complex number has n different complex nth roots, including the real ones (at most two). The nth root of 0 is zero for all positive integers n, since $0^n = 0$. In particular, if n is even and x is a positive real number, one of its nth roots is real and positive, one is negative, and the others (when n > 2) are non-real complex numbers; if n is even and x is a negative real number, none of the nth roots is real. If n is odd and x is real, one nth root is real and has the same sign as x, while the other (n – 1) roots are not real. Finally, if x is not real, then none of its nth roots are real.

Roots of real numbers are usually written using the radical symbol or radix $\sqrt{\ }$, with $\sqrt{x}$ denoting the positive square root of x if x is positive; for higher roots, $\sqrt[n]{x}$ denotes the real nth root if n is odd, and the positive nth root if n is even and x is positive. In the other cases, the symbol is not commonly used as being ambiguous. In the expression , the integer n is called the index and x is called the radicand.

When complex nth roots are considered, it is often useful to choose one of the roots, called principal root, as a principal value. The common choice is to choose the principal nth root of x as the nth root with the greatest real part, and when there are two (for x real and negative), the one with a positive imaginary part. This makes the nth root a function that is real and positive for x real and positive, and is continuous in the whole complex plane, except for values of x that are real and negative.

A difficulty with this choice is that, for a negative real number and an odd index, the principal nth root is not the real one. For example, "8 has three cube roots, "2 and The real cube root is "2 and the principal cube root is

An unresolved root, especially one using the radical symbol, is sometimes referred to as a surd or a radical. Any expression containing a radical, whether it is a square root, a cube root, or a higher root, is called a radical expression, and if it contains no transcendental functions or transcendental numbers it is called an algebraic expression.

Roots can also be defined as special cases of exponentiation, where the exponent is a fraction:

Roots are used for determining the radius of convergence of a power series with the root test. The nth roots of 1 are called roots of unity and play a fundamental role in various areas of mathematics, such as number theory, theory of equations, and Fourier transform.

Surds Definition

Surds are the square roots (") of numbers that cannot be simplified into a whole or rational number. It cannot be accurately represented in a fraction. In other words, a surd is a root of the whole number that has an irrational value. Consider an example, "2 H" 1.414213. It is more accurate if we leave it as a surd "2.

Types of Surds

There are six different types of surds, namely: Simple surds, Pure Surds, Similar Surds, Mixed Surds, Compound Surds, and Binomial Surds. Now let's understand these different types of surds"

- Simple Surds – A surd that has only one term is called a simple surd. Example: "2, "5, ...
- Pure Surds – Surds which are completely irrational. Example: "3
- Similar Surds – The surds having the same common surds factor
- Mixed Surds – Surds that are not completely irrational and can be expressed as a product of a rational number and an irrational number
- Compound Surds – An expression which is the addition or subtraction of two or more surds
- Binomial Surds – A surd that is made of two other surds

History

The Yale Babylonian Collection YBC 7289 clay tablet was created between 1800 BC and 1600 BC, showing and respectively as 1;24,51,10 and 0;42,25,35 base 60 numbers on a square crossed by two diagonals. (1;24,51,10) base 60 corresponds to 1.41421296, which is a correct value to 5 decimal points (1.41421356...).

The Rhind Mathematical Papyrus is a copy from 1650 BC of an earlier Berlin Papyrus and other texts – possibly the Kahun Papyrus – that shows how the Egyptians extracted square roots by an inverse proportion method.

In Ancient India, the knowledge of theoretical and applied aspects of square and square root was at least as old as the Sulba Sutras, dated around 800–500 BC (possibly much earlier). A method for finding very good approximations to the square roots of 2 and 3 are given in the Baudhayana Sulba Sutra. Aryabhata, in the Aryabhatiya (section 2.4), has given a method for finding the square root of numbers having many digits.

It was known to the ancient Greeks that square roots of positive integers that are not perfect squares are always irrational numbers: numbers not expressible as a ratio of two integers (that is, they cannot be written exactly as m/n, where m and n are integers). This is the theorem Euclid X, 9, almost certainly due to Theaetetus dating back to circa 380 BC. The particular case of the square root of 2 is assumed to date back earlier to the Pythagoreans, and is traditionally attributed to Hippasus. It is exactly the length of the diagonal of a square with side length 1.

In the Chinese mathematical work Writings on Reckoning, written between 202 BC and 186 BC during the early Han Dynasty, the square root is approximated by using an "excess and deficiency" method, which says to "...combine the excess and deficiency as the divisor; (taking) the deficiency numerator multiplied by the excess denominator and the excess numerator times the deficiency denominator, combine them as the dividend."

A symbol for square roots, written as an elaborate R, was invented by Regiomontanus (1436–1476). An R was also used for radix to indicate square roots in Gerolamo Cardano's Ars Magna.

According to historian of mathematics D.E. Smith, Aryabhata's method for finding the square root was first introduced in Europe by Cataneo—in 1546.

The symbol """ for the square root was first used in print in 1525, in Christoph Rudolff's Coss.

Surds and Indices

Surds are the root values that cannot be written as whole numbers. Indices are the power or exponent of a value. For example, for 32, 2 is the index and 3 is the base.

Suppose, we have a value "33. This value can be written as:

"$3^3 = (3^3)^{½} = 3^{3/2}$

Where 3/2 is the index.

Let us solve some examples here:

Example 1: $16^{3/2} + 16^{-3/2} = ?$

We know, by laws of exponents,

$a^m \times a^n = a^{m+n}$

$a^{-m} = 1/a^m$

⇒!$16^{3/2} + 1/16^{3/2}$

⇒!$(16^{1/2})^3 + 1/(16^{1/2})^3$

⇒!$(4^{2 \times 1/2})^3 + 1/(4^{2 \times 1/2})^3$

⇒!$4^3 + 1/4^3$

⇒! $64 + 1/64$

⇒! $(64 \times 64+ 1)/64$

$= (4096+1)/64$

$= 4097/64$

Example 2: If $(1/5\,)^{3a} = 0.008$ Find the value of $(\,0.25)^a$

Solution:

$(1/5\,)^{3a} = 0.008 = 8/1000 = 1/125 = (1/5\,)^3$

Apply the Algebra Law and Solve the equation.

⇒! 3^a = 3

4" a = 1

4" (0.25 $)^a$ = (0.25 $)^1$ = 0.25

How to Solve Surds?

You need to follow some rules to solve expressions that involve surds. One method is to rationalize the denominators, which helps to eject the surd in the denominator. Sometimes it may be mandatory to find the greatest perfect square factor to solve surds.

Problems and Solutions

Example 1: Write down the conjugate of 5"3 + ″2

Solution:

The conjugate of 5"3 + ″2 is 5"3 – ″2.

Example 2: Rationalise the denominator: 1/[(8"11)- (7"5)]

Solution:

Given: 1/[(8"11)- (7"5)]

It is known that the conjugate of (8"11)- (7"5) is (8"11)+(7"5)

To rationalize the denominator of the given fraction, multiply the conjugate of denominator on both numerator and denominator.

=[1/[(8"11)- (7"5)]]× [[(8"11)+ (7"5)]/[(8"11)+(7"5)]]

=[(8"11)+ (7"5)]/[(8"11 $)^2$-(7"5$)^2$]

=[(8"11)+ (7"5)]/[704- 245]

= [(8"11)+ (7"5)]/459

Example 3: Multiply ″7 x ″2

Solution: ″7 x ″2 = ″(7 x 2) = ″14

Example 4: Divide ″10 by ″5.

Solution: ″10/″5 = ″(10/5) = ″2

Example 5: Solve ″x + 2″x.

Solution: ″x + 2″x

= 3"x

Computation

Square roots of positive numbers are not in general rational numbers, and so cannot be written as a terminating or recurring decimal expression. Therefore in general any attempt to compute a square root expressed in decimal form can only yield an approximation, though a sequence of increasingly accurate approximations can be obtained.

Most pocket calculators have a square root key. Computer spreadsheets and other software are also frequently used to calculate square roots. Pocket calculators typically implement efficient routines, such as the Newton's method (frequently with an initial guess of 1), to compute the square root of a positive real number. When computing square roots with logarithm tables or slide rules, one can exploit the identities

where ln and $\log_{10}$ are the natural and base-10 logarithms.

By trial-and-error, one can square an estimate for and raise or lower the estimate until it agrees to sufficient accuracy. For this technique it is prudent to use the identity

as it allows one to adjust the estimate x by some amount c and measure the square of the adjustment in terms of the original estimate and its square. Furthermore, $(x + c)^2$ H" $x^2 + 2xc$ when c is close to 0, because the tangent line to the graph of $x^2 + 2xc + c^2$ at $c = 0$, as a function of c alone, is $y = 2xc + x^2$. Thus, small adjustments to x can be planned out by setting 2xc to a, or $c = a/(2x)$.

The most common iterative method of square root calculation by hand is known as the "Babylonian method" or "Heron's method" after the first-century Greek philosopher Heron of Alexandria, who first described it. The method uses the same iterative scheme as the Newton–Raphson method yields when applied to the function $y = f(x) = x^2$ " a, using the fact that its slope at any point is $dy/dx =$ f2 $(x) = 2x$, but predates it by many centuries. The algorithm is to repeat a simple calculation that results in a number closer to the actual square root each time it is repeated with its result as the new input.

The motivation is that if x is an overestimate to the square root of a nonnegative real number a then a/x will be an underestimate and so the average of these two numbers is a better approximation than either of them. However, the inequality of arithmetic and geometric means shows this average is always an overestimate of the square root (as noted below), and so it can serve as a new overestimate with which to repeat the process, which converges as a consequence of the successive overestimates and underestimates being closer to each other after each iteration. To find x:

1. Start with an arbitrary positive start value x. The closer to the square root of a, the fewer the iterations that will be needed to achieve the desired precision.
2. Replace x by the average $(x + a/x) / 2$ between x and a/x.
3. Repeat from step 2, using this average as the new value of x.

That is, if an arbitrary guess for is x_0, and $x_{n+1} = (x_n + a/x_n) / 2$, then each x_n is an approximation of which is better for large n than for small n. If a is positive, the convergence is quadratic, which means that in approaching the limit, the number of correct digits roughly doubles in each next iteration. If $a = 0$, the convergence is only linear.

USING THE IDENTITY

the computation of the square root of a positive number can be reduced to that of a number in the range [1,4). This simplifies finding a start value for the iterative method that is close to the square root, for which a polynomial or piecewise-linear approximation can be used.

The time complexity for computing a square root with n digits of precision is equivalent to that of multiplying two n-digit numbers.

Another useful method for calculating the square root is the shifting nth root algorithm, applied for n = 2.

The name of the square root function varies from programming language to programming language, with sqrt (often pronounced "squirt") being common, used in C, C++, and derived languages like JavaScript, PHP, and Python.

In integral domains, including fields

Each element of an integral domain has no more than 2 square roots. The difference of two squares identity u^2 " v^2 = (u " v)(u + v) is proved using the commutativity of multiplication. If u and v are square roots of the same element, then u^2 " v^2 = 0. Because there are no zero divisors this implies u = v or u + v = 0, where the latter means that two roots are additive inverses of each other. In other words if an element a square root u of an element a exists, then the only square roots of a are u and "u. The only square root of 0 in an integral domain is 0 itself.

In a field of characteristic 2, an element either has one square root or does not have any at all, because each element is its own additive inverse, so that "u = u. If the field is finite of characteristic 2 then every element has a unique square root. In a field of any other characteristic, any non-zero element either has two square roots, as explained above, or does not have any.

Given an odd prime number p, let $q = p^e$ for some positive integer e. A non-zero element of the field F_q with q elements is a quadratic residue if it has a square root in F_q. Otherwise, it is a quadratic non-residue. There are (q " 1)/2 quadratic residues and (q " 1)/2 quadratic non-residues; zero is not counted in either class. The quadratic residues form a group under multiplication. The properties of quadratic residues are widely used in number theory.

In rings in general

Unlike in an integral domain, a square root in an arbitrary (unital) ring need not be unique up to sign. For example, in the ring of integers modulo 8 (which is commutative, but has zero divisors), the element 1 has four distinct square roots: ±1 and ±3. Another example is provided by the ring of quaternions H, which has no zero divisors, but is not commutative. Here, the element "1 has infinitely many square roots, including ±i, ±j, and ±k. In fact, the set of square roots of "1 is exactly

A square root of 0 is either 0 or a zero divisor. Thus in rings where zero divisors do not exist, it is uniquely 0. However, rings with zero divisors may have multiple square roots of 0. For example, in any multiple of n is a square root of 0.

Geometric construction of the square root

The square root of a positive number is usually defined as the side length of a square with the area equal to the given number. But the square shape is not necessary for it: if one of two similar planar Euclidean objects has the area a times greater than another, then the ratio of their linear sizes is .

A square root can be constructed with a compass and straightedge. In his Elements, Euclid (fl. 300 BC) gave the construction of the geometric mean of two quantities in two different places: Proposition II.14 and Proposition VI.13. Since the geometric mean of a and b is , one can construct simply by taking b = 1.

The construction is also given by Descartes in his La Géométrie, see figure 2 on page 2. However, Descartes made no claim to originality and his audience would have been quite familiar with Euclid.

Euclid's second proof in Book VI depends on the theory of similar triangles. Let AHB be a line segment of length a + b with AH = a and HB = b. Construct the circle with AB as diameter and let C be one of the two intersections of the perpendicular chord at H with the circle and denote the length CH as h. Then, using Thales' theorem and, as in the proof of Pythagoras' theorem by similar triangles, triangle AHC is similar to triangle CHB (as indeed both are to triangle ACB, though we don't need that, but it is the essence of the proof of Pythagoras' theorem) so that AH:CH is as HC:HB, i.e. a/h = h/b, from which we conclude by cross-multiplication that $h^2 = ab$, and finally that . When marking the midpoint O of the line segment AB and drawing the radius OC of length (a + b)/2, then clearly OC > CH, i.e. (with equality if and only if a = b), which is the arithmetic–geometric mean inequality for two variables and, as noted above, is the basis of the Ancient Greek understanding of "Heron's method".

Another method of geometric construction uses right triangles and induction: can be constructed, and once has been constructed, the right triangle with legs 1 and has a hypotenuse of . Constructing successive square roots in this manner yields the Spiral of Theodorus depicted above.

METHODS OF COMPUTING SQUARE ROOTS

Methods of computing square roots are numerical analysis algorithms for approximating the principal, or non-negative, square root (usually denoted) of a real number. Arithmetically, it means given S, a procedure for finding a number which when multiplied by itself, yields S; algebraically, it means a procedure for finding the non-negative root of the equation x^2 " S = 0; geometrically, it means given two line segments, a procedure for constructing their geometric mean.

Every real number has two square roots. The principal square root of most numbers is an irrational number with an infinite decimal expansion. As a result, the decimal expansion of any such square root can only be computed to some finite-precision approximation. However, even if we are taking the square root of a perfect square integer, so that the result does have an exact finite representation, the procedure used to compute it may only return a series of increasingly accurate approximations.

The continued fraction representation of a real number can be used instead of its decimal or binary expansion and this representation has the property that the square root of any rational number (which is not already a perfect square) has a periodic, repeating expansion, similar to how rational numbers have repeating expansions in the decimal notation system.

The most common analytical methods are iterative and consist of two steps: finding a suitable starting value, followed by iterative refinement until some termination criterion is met. The starting value can be any number, but fewer iterations will be required the closer it is to the final result. The most familiar such method, most suited for programmatic calculation, is Newton's method, which is based on a property of the derivative in the calculus. A few methods like paper-and-pencil synthetic division and series expansion, do not require a starting value. In some applications, an integer square root is required, which is the square root rounded or truncated to the nearest integer (a modified procedure may be employed in this case).

The method employed depends on what the result is to be used for (i.e. how accurate it has to be), how much effort one is willing to put into the procedure, and what tools are at hand. The methods may be roughly classified as those suitable for mental calculation, those usually requiring at least paper and pencil, and those which are implemented as programs to be executed on a digital electronic computer or other computing device. Algorithms may take into account convergence (how many iterations are required to achieve a specified precision), computational complexity of individual operations (i.e. division) or iterations, and error propagation (the accuracy of the final result).

Procedures for finding square roots (particularly the square root of 2) have been known since at least the period of ancient Babylon in the 17th century BCE. Heron's method from first century Egypt was the first ascertainable algorithm for computing square root. Modern analytic methods began to be developed after introduction of the Arabic numeral system to western Europe in the early Renaissance. Today, nearly all computing devices have a fast and accurate square root function, either as a programming language construct, a compiler intrinsic or library function, or as a hardware operator, based on one of the described procedures.

INITIAL ESTIMATE

Many iterative square root algorithms require an initial seed value. The seed must be a non-zero positive number; it should be between 1 and S, the number whose square root is desired, because the square root must be in that range. If the seed is far away from the root, the algorithm will require more iterations. If one initializes with $x_0 = 1$ (or S), then approximately iterations will be wasted just getting the order of magnitude of the root. It is therefore useful to have a rough estimate, which may have limited accuracy but is easy to calculate. In general, the better the initial estimate, the faster the convergence. For Newton's method (also called Babylonian or Heron's method), a seed somewhat larger than the root will converge slightly faster than a seed somewhat smaller than the root.

In general, an estimate is pursuant to an arbitrary interval known to contain the root (such as). The estimate is a specific value of a functional approximation to over the interval. Obtaining a better estimate involves either obtaining tighter bounds on the interval, or finding a better functional approximation to f(x). The latter usually means using a higher order polynomial in the approximation, though not all approximations are polynomial. Common methods of estimating include scalar, linear, hyperbolic and logarithmic. A decimal base is usually used for mental or paper-and-pencil estimating. A binary base is more suitable for computer estimates. In estimating, the exponent and mantissa are usually treated separately, as the number would be expressed in scientific notation.

Decimal estimates

Typically the number S is expressed in scientific notation as $a \times 10^{2n}$ where 1 d" a < 100 and n is an integer, and the range of possible square roots is $a \times 10^{n}$ where 1 d" a < 10.

Scalar estimates

Scalar methods divide the range into intervals, and the estimate in each interval is represented by a single scalar number. If the range is cosidered as a single interval, the arithmetic mean (5.5) or geometric mean () times 10^n are plausible estimates. The absolute and relative error for these will differ. In general, a single scalar will be very inaccurate. Better estimates divide the range into two or more intervals, but scalar estimates have inherently low accuracy.

For two intervals, divided geometrically, the square root can be estimated as

This estimate has maximum absolute error of 4 Å" 10^n at a = 100, and maximum relative error of 100% at a = 1.

For example, for S = 125348 factored as 12.5348×10^4, the estimate is . , an absolute error of 246 and relative error of almost 70%.

Linear estimates

A better estimate, and the standard method used, is a linear approximation to the function $y = x^2$ over a small arc. If, as above, powers of the base are factored out of the number S and the interval reduced to [1, 100], a secant line spanning the arc, or a tangent line somewhere along the arc may be used as the approximation, but a least-squares regression line intersecting the arc will be more accurate.

A least-squares regression line minimizes the average difference between the estimate and the value of the function. Its equation is y = 8.7x " 10. Reordering, x = 0.115 y + 1.15. Rounding the coefficients for ease of computation,

That is the best estimate on average that can be achieved with a single piece linear approximation of the function $y = x^2$ in the interval [1, 100] . It has a maximum absolute error of 1.2 at a=100, and maximum relative error of 30% at S=1 and 10.

To divide by 10, subtract one from the exponent of a, or figuratively move the decimal point one digit to the left. For this formulation, any additive constant 1 plus a small increment will make a satisfactory estimate so remembering the exact number isn't a burden. The approximation (rounded or not) using a single line spanning the range [1, 100] is less than one significant digit of precision; the relative error is greater than $1/2^2$, so less than 2 bits of information are provided. The accuracy is severely limited because the range is two orders of magnitude, quite large for this kind of estimation.

A much better estimate can be obtained by a piece-wise linear approximation: multiple line segments, each approximating some subarc of the original. The more line segments used, the better the approximation. The most common way is to use tangent lines; the critical choices are how to divide the arc and where to place the tangent points.

An efficacious way to divide the arc from y=1 to y=100 is geometrically: for two intervals, the bounds of the intervals are the square root of the bounds of the original interval, 1*100, i.e. [1,^{2}"100] and [^{2}"100,100]. For three intervals, the bounds are the cube roots of 100: [1,^{3}"100], [^{3}"100,(^{3}"100)^{2}], and [(^{3}"100)^{2},100], etc. For two intervals, ^{2}"100 = 10, a very convenient number. Tangent lines are easy to derive, and are located at x = "1*"10 and x = "10*"10. Their equations are: y = 3.56 x " 3.16 and y = 11.2 x " 31.6 . Inverting, the square roots are: x = 0.28 y + 0.89 and x = .089 y + 2.8. Thus for :

The maximum absolute errors occur at the high points of the intervals, at a=10 and 100, and are 0.54 and 1.7 respectively. The maximum relative errors are at the endpoints of the intervals, at a=1, 10 and 100, and are 17% in both cases. 17% or 0.17 is larger than 1/10, so the method yields less than a decimal digit of accuracy.

Hyperbolic estimates

In some cases, hyperbolic estimates may be efficacious, because a hyperbola is also a convex curve and may lie along an arc of $Y = x^2$ better than a line. Hyperbolic estimates are more computationally complex, because they necessarily require a floating division. A near-optimal hyperbolic approximation to x2 on the interval [1, 100] is y=190/(10-x)-20. Transposing, the square root is x = -190/(y+20) + 10. Thus for :

The floating division need be accurate to only one decimal digit, because the estimate overall is only that accurate, and can be done mentally. A hyperbolic estimate is better on average than scalar or linear estimates. It has maximum absolute error of 1.58 at 100 and maximum relative error of 16.0% at 10. For the worst case at a=10, the estimate is 3.67. If one starts with 10 and applies Newton-Raphson iterations straight away, two iterations will be required, yielding 3.66, before the accuracy of the hyperbolic estimate is exceeded. For a more typical case like 75, the hyperbolic estimate is 8.00, and 5 Newton-Raphson iterations starting at 75 would be required to obtain a more accurate result.

Arithmetic estimates

A method analogous to piece-wise linear approximation but using only arithmetic instead of algebraic equations, uses the multiplication tables in reverse: the square root of a number between 1 and 100 is between 1 and 10, so if we know 25 is a perfect square (5 × 5), and 36 is a perfect square (6 × 6), then the square root of a number greater than or equal to 25 but less than 36, begins with a 5. Similarly for numbers between other squares. This method will yield a correct first digit, but it is not accurate to one digit: the first digit of the square root of 35 for example, is 5, but the square root of 35 is almost 6.

A better way is to the divide the range into intervals half way between the squares. So any number between 25 and half way to 36, which is 30.5, estimate 5; any number greater than 30.5 up to 36, estimate 6.[Note 4] The procedure only requires a little arithmetic to find a boundary number in the middle of two products from the multiplication table. Here is a reference table of those boundaries:

a	**nearest square**	$k = \sqrt{a}$ **est.**
1 to 2.5	1 (= 1^2)	1
2.5 to 6.5	4 (= 2^2)	2
6.5 to 12.5	9 (= 3^2)	3
12.5 to 20.5	16 (= 4^2)	4
20.5 to 30.5	25 (= 5^2)	5
30.5 to 42.5	36 (= 6^2)	6
42.5 to 56.5	49 (= 7^2)	7
56.5 to 72.5	64 (= 8^2)	8
72.5 to 90.5	81 (= 9^2)	9
90.5 to 100	100 (= 10^2)	10

Digit-by-digit calculation

This is a method to find each digit of the square root in a sequence. It is slower than the Babylonian method, but it has several advantages:

- It can be easier for manual calculations.
- Every digit of the root found is known to be correct, i.e., it does not have to be changed later.
- If the square root has an expansion that terminates, the algorithm terminates after the last digit is found. Thus, it can be used to check whether a given integer is a square number.
- The algorithm works for any base, and naturally, the way it proceeds depends on the base chosen.

Napier's bones include an aid for the execution of this algorithm. The shifting nth root algorithm is a generalization of this method.

Decimal (base 10)

Write the original number in decimal form. The numbers are written similar to the long division algorithm, and, as in long division, the root will be written on the line above. Now separate the digits into pairs, starting from the decimal point and going both left and right. The decimal point of the root will be above the decimal point of the square. One digit of the root will appear above each pair of digits of the square.

Beginning with the left-most pair of digits, do the following procedure for each pair:

1. Starting on the left, bring down the most significant (leftmost) pair of digits not yet used (if all the digits have been used, write "00") and write them to the right of the remainder from the previous step (on the first step, there will be no remainder). In other words, multiply the remainder by 100 and add the two digits. This will be the current value c.
2. Find p, y and x, as follows:
   - o Let p be the part of the root found so far, ignoring any decimal point. (For the first step, p = 0.)
   - o Determine the greatest digit x such that x (20 p + x) d" c. We will use a new variable y = x(20p + x).
   - \- Note: 20p + x is simply twice p, with the digit x appended to the right.
   - \- Note: x can be found by guessing what c/(20·p) is and doing a trial calculation of y, then adjusting x upward or downward as necessary.
   - o Place the digit x as the next digit of the root, i.e., above the two digits of the square you just brought down. Thus the next p will be the old p times 10 plus x.
3. Subtract y from c to form a new remainder.
4. If the remainder is zero and there are no more digits to bring down, then the algorithm has terminated. Otherwise go back to step 1 for another iteration.

Approximations that depend on the floating point representation

A number is represented in a floating point format as $m \times b^p$ which is also called scientific notation. Its square root is $\sqrt{m} \times b^{p/2}$ and similar formulae would apply for cube roots and logarithms. On the face of it, this is no improvement in simplicity, but suppose that only an approximation is required: then just $b^{p/2}$ is good to an order of magnitude. Next, recognise that some powers, p, will be odd, thus for $3141.59 = 3.14159\times10^3$ rather than deal with fractional powers of the base, multiply the mantissa by the base and subtract one from the power to make it even. The adjusted representation will become the equivalent of 31.4159×10^2 so that the square root will be "31.4159×10^1.

If the integer part of the adjusted mantissa is taken, there can only be the values 1 to 99, and that could be used as an index into a table of 99 pre-computed square roots to complete the estimate. A computer using base sixteen would require a larger table, but one using base two would require only three entries: the possible bits of the integer part of the adjusted mantissa are 01 (the power being even so there was no shift, remembering that a normalised floating point number always has a non-zero high-order digit) or if the power was odd, 10 or 11, these being the first two bits of the original mantissa.

Thus, 6.25 = 110.01 in binary, normalised to 1.1001×2^2 an even power so the paired bits of the mantissa are 01, while .625 = 0.101 in binary normalises to $1.01 \times 2^{''1}$ an odd power so the adjustment is to $10.1 \times 2^{''2}$ and the paired bits are 10. Notice that the low order bit of the power is echoed in the high order bit of the pairwise mantissa. An even power has its low-order bit zero and the adjusted mantissa will start with 0, whereas for an odd power that bit is one and the adjusted mantissa will start with 1. Thus, when the power is halved, it is as if its low order bit is shifted out to become the first bit of the pairwise mantissa.

A table with only three entries could be enlarged by incorporating additional bits of the mantissa. However, with computers, rather than calculate an interpolation into a table, it is often better to find some simpler calculation giving equivalent results. Everything now depends on the exact details of the format of the representation, plus what operations are available to access and manipulate the parts of the number. For example, Fortran offers an EXPONENT(x) function to obtain the power. Effort expended in devising a good initial approximation is to be recouped by thereby avoiding the additional iterations of the refinement process that would have been needed for a poor approximation. Since these are few (one iteration requires a divide, an add, and a halving) the constraint is severe.

Many computers follow the IEEE (or sufficiently similar) representation, and a very rapid approximation to the square root can be obtained for starting Newton's method. The technique that follows is based on the fact that the floating point format (in base two) approximates the base-2 logarithm. That is log 2 (m × 2p) = p + log 2 (m).

So for a 32-bit single precision floating point number in IEEE format (where notably, the power has a bias of 127 added for the represented form) you can get the approximate logarithm by interpreting its binary representation as a 32-bit integer, scaling it by $2^{''23}$, and removing a bias of 127, i.e.

$$x_{\text{int}} \cdot 2^{-23} - 127 \approx \log_2(x).$$

Matrices with distinct eigenvalues

An n×n matrix with n distinct nonzero eigenvalues has 2^n square roots. Such a matrix, A, has an eigendecomposition $VDV^{''1}$ where V is the matrix whose columns are eigenvectors of A and D is the diagonal matrix whose diagonal elements are the corresponding n eigenvalues $\ddot{e}_i$. Thus the square roots of A are given by $VD^{1/2}V^{"1}$, where $D^{1/2}$ is any square root matrix of D, which, for distinct eigenvalues, must be diagonal with diagonal elements equal to square roots of the diagonal elements of D; since there are two possible choices for a square root of each diagonal element of D, there are 2^n choices for the matrix $D^{1/2}$.

This also leads to a proof of the above observation, that a positive-definite matrix has precisely one positive-definite square root: a positive definite matrix has only positive eigenvalues, and each of these eigenvalues has only one positive square root; and since the eigenvalues of the square root matrix are the diagonal elements of $D^{1/2}$, for the square root matrix to be itself positive definite necessitates the use of only the unique positive square roots of the original eigenvalues.

Square roots of positive operators

In linear algebra and operator theory, given a bounded positive semidefinite operator (a non-negative operator) T on a complex Hilbert space, B is a square root of T if T = B* B, where B* denotes the Hermitian adjoint of B. According to the spectral theorem, the continuous functional calculus can be applied to obtain an operator $T^{1/2}$ such that $T^{1/2}$ is itself positive and $(T^{1/2})^2 = T$. The operator $T^{1/2}$ is the unique non-negative square root of T.

A bounded non-negative operator on a complex Hilbert space is self adjoint by definition. So $T = (T^{1/2})^* T^{1/2}$. Conversely, it is trivially true that every operator of the form B* B is non-negative. Therefore, an operator T is non-negative if and only if T = B* B for some B (equivalently, T = CC* for some C).

The Cholesky factorization provides another particular example of square root, which should not be confused with the unique non-negative square root.

Unitary freedom of square roots

If T is a non-negative operator on a finite-dimensional Hilbert space, then all square roots of T are related by unitary transformations. More precisely, if T = A*A = B*B, then there exists a unitary U such that A = UB.

Indeed, take $B = T^{1/2}$ to be the unique non-negative square root of T. If T is strictly positive, then B is invertible, and so $U = AB^{"1}$ is unitary:

$$U^*U = \left(\left(B^*\right)^{-1} A^*\right)\left(AB^{-1}\right) = \left(B^*\right)^{-1} T\left(B^{-1}\right)$$

$$= \left(B^*\right)^{-1} B^* B\left(B^{-1}\right) = I.$$

If T is non-negative without being strictly positive, then the inverse of B cannot be defined, but the Moore–Penrose pseudoinverse B^+ can be. In that case, the operator B^+A is a partial isometry, that is, a unitary operator from the range of T to itself. This can then be extended to a unitary operator U on the whole space by setting it equal to the identity on the kernel of T. More generally, this is true on an infinite-dimensional Hilbert space if, in addition, T has closed range. In general, if A, B are closed and densely defined operators on a Hilbert space H, and $A^* A = B^* B$, then A = UB where U is a partial isometry.

Some applications

Square roots, and the unitary freedom of square roots, have applications throughout functional analysis and linear algebra.

Polar decomposition

If A is an invertible operator on a finite-dimensional Hilbert space, then there is a unique unitary operator U and positive operator P such that

A = U P;

this is the polar decomposition of A. The positive operator P is the unique positive square root of the positive operator A″A, and U is defined by U = $AP^{''1}$.

If A is not invertible, then it still has a polar composition in which P is defined in the same way (and is unique). The unitary operator U is not unique. Rather it is possible to determine a "natural" unitary operator as follows: AP^+ is a unitary operator from the range of A to itself, which can be extended by the identity on the kernel of A″. The resulting unitary operator U then yields the polar decomposition of A.

9

Multiplication and Division of Integers

The arithmetic operation is a branch of Mathematics that involves addition, subtraction, division and multiplication on all types of real numbers including integers.

Integers are a particular kind of numbers that include negative numbers, positive numbers and zero and no fractions. Arithmetic operations on integers and real numbers are similar that can be either positive or negative. Rules for multiplication and division of integers are explained below.

MULTIPLICATION OF INTEGERS

Multiplication (often denoted by the cross symbol ×, by the mid-line dot operator Å", by juxtaposition, or, on computers, by an asterisk *) is one of the four elementary mathematical operations of arithmetic, with the other ones being addition, subtraction, and division. The result of a multiplication operation is called a product.

The multiplication of whole numbers may be thought of as repeated addition; that is, the multiplication of two numbers is equivalent to adding as many copies of one of them, the multiplicand, as the quantity of the other one, the multiplier. Both numbers can be referred to as factors.

$$a \times b = \underbrace{b + \cdots + b}_{a \text{ times}}$$

For example, 4 multiplied by 3, often written as 3 × 4 and spoken as "3 times 4", can be calculated by adding 3 copies of 4 together:

$3 \times 4 = 4 + 4 + 4 = 12$

Here, 3 (the multiplier) and 4 (the multiplicand) are the factors, and 12 is the product.

One of the main properties of multiplication is the commutative property, which states in this case that adding 3 copies of 4 gives the same result as adding 4 copies of 3:

$4 \times 3 = 3 + 3 + 3 + 3 = 12$

Thus the designation of multiplier and multiplicand does not affect the result of the multiplication.

Systematic generalizations of this basic definition define the multiplication of integers (including negative numbers), rational numbers (fractions), and real numbers.

Multiplication can also be visualized as counting objects arranged in a rectangle (for whole numbers) or as finding the area of a rectangle whose sides have some given lengths. The area of a rectangle does not depend on which side is measured first—a consequence of the commutative property.

The product of two measurements is a new type of measurement. For example, multiplying the lengths of the two sides of a rectangle gives its area. Such a product is the subject of dimensional analysis.

The inverse operation of multiplication is division. For example, since 4 multiplied by 3 equals 12, 12 divided by 3 equals 4. Indeed, multiplication by 3, followed by division by 3, yields the original number. The division of a number other than 0 by itself equals 1.

Multiplication is also defined for other types of numbers, such as complex numbers, and for more abstract constructs, like matrices. For some of these more abstract constructs, the order in which the operands are multiplied together matters. A listing of the many different kinds of products used in mathematics is given in Product (mathematics).

In arithmetic, multiplication is often written using the multiplication sign (either × or ×) between the terms (that is, in infix notation). For example,

$2 \times 3 = 6$ ("two times three equals six")

$3 \times 4 = 12$

$2 \times 3 \times 5 = 6 \times 5 = 30$

$2 \times 2 \times 2 \times 2 \times 2 = 32$

There are other mathematical notations for multiplication:

- To reduce confusion between the multiplication sign × and the common variable x, multiplication is also denoted by dot signs, usually a middle-position dot (rarely period):
  5 Å" 2 or 5 . 3

The middle dot notation, encoded in Unicode as U+22C5 Å" DOT OPERATOR, is now standard in the United States and other countries where the period is used as a decimal point. When the dot operator character is not accessible, the interpunct (·) is used. In other countries that use a comma as a decimal mark, either the period or a middle dot is used for multiplication.

Historically, in the United Kingdom and Ireland, the middle dot was sometimes used for the decimal to prevent it from disappearing in the ruled line, and the period/full stop was used for multiplication. However, since the Ministry of Technology ruled to use the period as the decimal point in 1968, and the SI standard has since been widely adopted, this usage is now found only in the more traditional journals such as The Lancet.

- In algebra, multiplication involving variables is often written as a juxtaposition (e.g., xy for x times y or 5x for five times x), also called implied multiplication. The notation can also be used for quantities that are surrounded by parentheses (e.g., 5(2) or (5)(2) for five times two). This implicit usage of multiplication can cause ambiguity when the concatenated variables happen to match the name of another variable, when a variable name in front of a parenthesis can be confused with a function name, or in the correct determination of the order of operations.
- In vector multiplication, there is a distinction between the cross and the dot symbols. The cross symbol generally denotes the taking a cross product of two vectors, yielding a vector as its result, while the dot denotes taking the dot product of two vectors, resulting in a scalar.

In computer programming, the asterisk (as in 5*2) is still the most common notation. This is due to the fact that most computers historically were limited to small character sets (such as ASCII and EBCDIC) that lacked a multiplication sign (such as Å" or ×), while the asterisk appeared on every keyboard. This usage originated in the FORTRAN programming language.

The numbers to be multiplied are generally called the "factors". The number to be multiplied is the "multiplicand", and the number by which it is multiplied is the "multiplier". Usually, the multiplier is placed first and the multiplicand is placed second; however sometimes the first factor is the multiplicand and the second the multiplier. Also, as the result of multiplication does not depend on the order of the factors, the distinction between "multiplicand" and "multiplier" is useful only at a very elementary level and in some multiplication algorithms, such as the long multiplication. Therefore, in some sources, the term "multiplicand" is regarded as a synonym for "factor". In algebra, a number that is the multiplier of a variable or expression (e.g., the 3 in $3xy^2$) is called a coefficient.

The result of a multiplication is called a product. When one factor is an integer, the product is a multiple of the other or of the product of the others. Thus $2 \times \pi$ is a multiple of π, as is $5133 \times 486 \times \pi$. A product of integers is a multiple of each factor; for example, 15 is the product of 3 and 5 and is both a multiple of 3 and a multiple of 5.

By definition, multiplication is the repeated addition of numbers. But the rules for multiplication of integers are different from that of addition. It includes three possibilities. They are:

- Multiplication between two positive numbers,
- Multiplication between two negative numbers; and
- Multiplication between a positive number and a negative number.

The product of two integers with similar sign numbers will always be positive. This means the product of two positive numbers or two negative numbers will always be positive. While the product of a positive number and negative number (integers with different signs) will always be negative.

Rules of Multiplication

Type of Numbers	Operation	Result	Example
Positive x Positive	Multiply	Positive (+)	1 x 5 = 5
Negative x Negative	Multiply	Positive (+)	(-1) x (-5) = 5
Positive x Negative	Multiply	Negative (-)	1 x (-5) = -5
Negative x Positive	Multiply	Negative (-)	(-1) x 5 = -5

In case of multiplication of integers, just multiply the numbers without the sign. Once the product is obtained give the sign according to the rule of multiplication given in the above table.

Note:

- If number of negative integers to multiply is odd, the product will be negative
- If number of negative integers to multiply is even, the product will be positive

Computation

Many common methods for multiplying numbers using pencil and paper require a multiplication table of memorized or consulted products of small numbers (typically any two numbers from 0 to 9). However, one method, the peasant multiplication algorithm, does not. The example below illustrates "long multiplication" (the "standard algorithm", "grade-school multiplication"):

```
        23958233
  ×         5830
  ——————————————
        00000000 ( =      23,958,233 ×      0)
       71874699  ( =      23,958,233 ×     30)
      191665864  ( =      23,958,233 ×    800)
  + 119791165    ( =      23,958,233 × 5,000)
  ——————————————
    139676498390 ( = 139,676,498,390         )
```

In some countries such as Germany, the above multiplication is depicted similarly but with the original product kept horizontal and computation starting with the first digit of the multiplier:

```
23958233 · 5830
───────────────
  119791165
   191665864
     71874699
      00000000
───────────────
  139676498390
```

Multiplying numbers to more than a couple of decimal places by hand is tedious and error-prone. Common logarithms were invented to simplify such calculations, since adding logarithms is equivalent to multiplying. The slide rule allowed numbers to be quickly multiplied to about three places of accuracy. Beginning in the early 20th century, mechanical calculators, such as the Marchant, automated multiplication of up to 10-digit numbers. Modern electronic computers and calculators have greatly reduced the need for multiplication by hand.

Historical algorithms

Methods of multiplication were documented in the writings of ancient Egyptian, Greek, Indian, and Chinese civilizations.

The Ishango bone, dated to about 18,000 to 20,000 BC, may hint at a knowledge of multiplication in the Upper Paleolithic era in Central Africa, but this is speculative.

Egyptians

The Egyptian method of multiplication of integers and fractions, which is documented in the Rhind Mathematical Papyrus, was by successive additions and doubling. For instance, to find the product of 13 and 21 one had to double 21 three times, obtaining $2 \times 21 = 42$, $4 \times 21 = 2 \times 42 = 84$, $8 \times 21 = 2 \times 84 = 168$. The full product could then be found by adding the appropriate terms found in the doubling sequence:

$13 \times 21 = (1 + 4 + 8) \times 21 = (1 \times 21) + (4 \times 21) + (8 \times 21) = 21 + 84 + 168 = 273$.

Babylonians

The Babylonians used a sexagesimal positional number system, analogous to the modern-day decimal system. Thus, Babylonian multiplication was very similar to modern decimal multiplication. Because of the relative difficulty of remembering 60×60 different products,

Babylonian mathematicians employed multiplication tables. These tables consisted of a list of the first twenty multiples of a certain principal number n: n, 2n, ..., 20n; followed by the multiples of 10n: 30n 40n, and 50n. Then to compute any sexagesimal product, say 53n, one only needed to add 50n and 3n computed from the table.

Chinese

In the mathematical text Zhoubi Suanjing, dated prior to 300 BC, and the Nine Chapters on the Mathematical Art, multiplication calculations were written out in words, although the early Chinese mathematicians employed Rod calculus involving place value addition, subtraction, multiplication, and division. The Chinese were already using a decimal multiplication table by the end of the Warring States period.

Modern methods

The modern method of multiplication based on the Hindu–Arabic numeral system was first described by Brahmagupta. Brahmagupta gave rules for addition, subtraction, multiplication, and division. Henry Burchard Fine, then a professor of mathematics at Princeton University, wrote the following:

The Indians are the inventors not only of the positional decimal system itself, but of most of the processes involved in elementary reckoning with the system. Addition and subtraction they performed quite as they are performed nowadays; multiplication they effected in many ways, ours among them, but division they did cumbrously.

These place value decimal arithmetic algorithms were introduced to Arab countries by Al Khwarizmi in the early 9th century and popularized in the Western world by Fibonacci in the 13th century.

Grid method

Grid method multiplication, or the box method, is used in primary schools in England and Wales and in some areas of the United States to help teach an understanding of how multiple digit multiplication works. An example of multiplying 34 by 13 would be to lay the numbers out in a grid as follows:

×	30	4
10	300	40
3	90	12

and then add the entries.

Computer algorithms

The classical method of multiplying two n-digit numbers requires n^2 digit multiplications. Multiplication algorithms have been designed that reduce the computation time considerably when multiplying large numbers. Methods based on the discrete Fourier transform reduce the computational complexity to O(n log n log log n). In 2016, the factor log log n was replaced by a function that increases much slower, though still not constant. In March 2019, David Harvey and Joris van der Hoeven submitted a paper presenting an integer multiplication algorithm with a complexity of O (n log n). The algorithm, also based on the fast Fourier transform, is conjectured to be asymptotically optimal. The algorithm is not practically useful, as it only becomes faster for multiplying extremely large numbers (having more than 2^{172912} bits).

PRODUCTS OF MEASUREMENTS

One can only meaningfully add or subtract quantities of the same type, but quantities of different types can be multiplied or divided without problems. For example, four bags with three marbles each can be thought of as:

[4 bags] × [3 marbles per bag] = 12 marbles.

When two measurements are multiplied together, the product is of a type depending on the types of measurements. The general theory is given by dimensional analysis. This analysis is routinely applied in physics, but it also has applications in finance and other applied fields.

A common example in physics is the fact that multiplying speed by time gives distance. For example:

50 kilometers per hour × 3 hours = 150 kilometers.

In this case, the hour units cancel out, leaving the product with only kilometer units.

Other examples of multiplication involving units include:

2.5 meters × 4.5 meters = 11.25 square meters

11 meters/seconds × 9 seconds = 99 meters

4.5 residents per house × 20 houses = 90 residents

Multiplication with set theory

The product of non-negative integers can be defined with set theory using cardinal numbers or the Peano axioms. See below how to extend this to multiplying arbitrary integers, and then arbitrary rational numbers. The product of real numbers is defined in terms of products of rational numbers; see construction of the real numbers.

Multiplication in group theory

There are many sets that, under the operation of multiplication, satisfy the axioms that define group structure. These axioms are closure, associativity, and the inclusion of an identity element and inverses.

A simple example is the set of non-zero rational numbers. Here we have identity 1, as opposed to groups under addition where the identity is typically 0. Note that with the rationals, we must exclude zero because, under multiplication, it does not have an inverse: there is no rational number that can be multiplied by zero to result in 1. In this example, we have an abelian group, but that is not always the case.

To see this, consider the set of invertible square matrices of a given dimension over a given field. Here, it is straightforward to verify closure, associativity, and inclusion of identity (the identity matrix) and inverses. However, matrix multiplication is not commutative, which shows that this group is non-abelian.

Another fact worth noticing is that the integers under multiplication do not form a group—even if we exclude zero. This is easily seen by the nonexistence of an inverse for all elements other than 1 and ″1.

Multiplication in group theory is typically notated either by a dot or by juxtaposition (the omission of an operation symbol between elements). So multiplying element a by element b could be notated as a Å″ b or ab. When referring to a group via the indication of the set and operation, the dot is used. For example, our first example could be indicated by (Q/{0}, Å″).

Multiplication algorithm

A multiplication algorithm is an algorithm (or method) to multiply two numbers. Depending on the size of the numbers, different algorithms are used. Efficient multiplication algorithms have existed since the advent of the decimal system.

Long multiplication

If a positional numeral system is used, a natural way of multiplying numbers is taught in schools as long multiplication, sometimes called grade-school multiplication, sometimes called Standard Algorithm: multiply the multiplicand by each digit of the multiplier and then add up all the properly shifted results. It requires memorization of the multiplication table for single digits.

This is the usual algorithm for multiplying larger numbers by hand in base 10. A person doing long multiplication on paper will write down all the products and then add them together; an abacus-user will sum the products as soon as each one is computed.

Example

This example uses long multiplication to multiply 23,958,233 (multiplicand) by 5,830 (multiplier) and arrives at 139,676,498,390 for the result (product).

```
        23958233
×           5830
————————————————
        00000000  ( =      23,958,233 ×     0)
       71874699   ( =      23,958,233 ×    30)
     191665864    ( =      23,958,233 ×   800)
+   119791165     ( =      23,958,233 × 5,000)
————————————————
    139676498390  ( = 139,676,498,390        )
```

Other notations

In some countries such as Germany, the above multiplication is depicted similarly but with the original product kept horizontal and computation starting with the first digit of the multiplier:

```
23958233 · 5830
———————————————
119791165
 191665864
   71874699
    00000000
———————————————
139676498390
```

Below pseudocode describes the process of above multiplication. It keeps only one row to maintain the sum which finally becomes the result. Note that the '+=' operator is used to denote sum to existing value and store operation (akin to languages such as Java and C) for compactness.

Usage in computers

Some chips implement long multiplication, in hardware or in microcode, for various integer and floating-point word sizes. In arbitrary-precision arithmetic, it is common to use long multiplication with the base set to 2^w, where w is the number of bits in a word, for multiplying relatively small numbers. To multiply two numbers with n digits using this method, one needs about n^2 operations. More formally, multiplying two n-digit numbers using long multiplication requires È(n^2) single-digit operations (additions and multiplications).

When implemented in software, long multiplication algorithms must deal with overflow during additions, which can be expensive. A typical solution is to represent the number in a small base, b, such that, for example,

8b is a representable machine integer. Several additions can then be performed before an overflow occurs. When the number becomes too large, we add part of it to the result, or we carry and map the remaining part back to a number that is less than b. This process is called normalization. Richard Brent used this approach in his Fortran package, MP.

Computers initially used a very similar algorithm to long multiplication in base 2, but modern processors have optimized circuitry for fast multiplications using more efficient algorithms, at the price of a more complex hardware realization. In base two, long multiplication is sometimes called "shift and add", because the algorithm simplifies and just consists of shifting left (multiplying by powers of two) and adding. Most currently available microprocessors implement this or other similar algorithms (such as Booth encoding) for various integer and floating-point sizes in hardware multipliers or in microcode.

On currently available processors, a bit-wise shift instruction is faster than a multiply instruction and can be used to multiply (shift left) and divide (shift right) by powers of two. Multiplication by a constant and division by a constant can be implemented using a sequence of shifts and adds or subtracts. For example, there are several ways to multiply by 10 using only bit-shift and addition.

```
((x << 2) + x) << 1 # Here 10*x is computed as (x*2^2 + x)*2
(x << 3) + (x << 1) # Here 10*x is computed as x*2^3 + x*2
```

In some cases such sequences of shifts and adds or subtracts will outperform hardware multipliers and especially dividers. A division by a number of the form 2^n or $2^n \pm 1$ often can be converted to such a short sequence.

Algorithms for multiplying by hand

In addition to the standard long multiplication, there are several other methods used to perform multiplication by hand. Such algorithms may be devised for speed, ease of calculation, or educational value, particularly when computers or multiplication tables are unavailable.

Grid method

The grid method (or box method) is an introductory method for multiple-digit multiplication that is often taught to pupils at primary school or elementary school. It has been a standard part of the national primary school mathematics curriculum in England and Wales since the late 1990s.

Both factors are broken up ("partitioned") into their hundreds, tens and units parts, and the products of the parts are then calculated explicitly in a relatively simple multiplication-only stage, before these contributions are then totalled to give the final answer in a separate addition stage.

The calculation 34 × 13, for example, could be computed using the grid:

```
300
   40
   90
+  12
-----
  442
```

×	30	4
10	300	40
3	90	12

followed by addition to obtain 442, either in a single sum, or through forming the row-by-row totals (300 + 40) + (90 + 12) = 340 + 102 = 442.

This calculation approach (though not necessarily with the explicit grid arrangement) is also known as the partial products algorithm. Its essence is the calculation of the simple multiplications separately, with all addition being left to the final gathering-up stage.

The grid method can in principle be applied to factors of any size, although the number of sub-products becomes cumbersome as the number of digits increases. Nevertheless, it is seen as a usefully explicit method to introduce the idea of multiple-digit multiplications; and, in an age when most multiplication calculations are done using a calculator or a spreadsheet, it may in practice be the only multiplication algorithm that some students will ever need.

Lattice multiplication

First, set up the grid by marking its rows and columns with the numbers to be multiplied. Then, fill in the boxes with tens digits in the top triangles and units digits on the bottom.

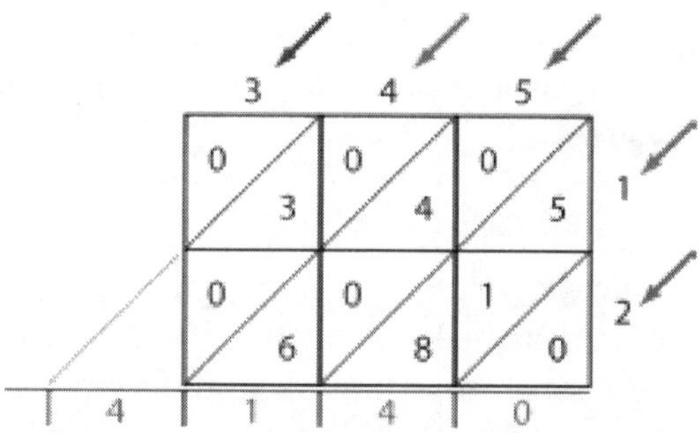

Finally, sum along the diagonal tracts and carry as needed to get the answer

Lattice, or sieve, multiplication is algorithmically equivalent to long multiplication. It requires the preparation of a lattice (a grid drawn on paper) which guides the calculation and separates all the multiplications from the additions. It was introduced to Europe in 1202 in Fibonacci's Liber Abaci. Fibonacci described the operation as mental, using his right and left hands to carry the intermediate calculations. Matrakçý Nasuh presented 6 different variants of this method in this 16th-century book, Umdet-ul Hisab. It was widely used in Enderun schools across the Ottoman Empire. Napier's bones, or Napier's rods also used this method, as published by Napier in 1617, the year of his death.

As shown in the example, the multiplicand and multiplier are written above and to the right of a lattice, or a sieve. It is found in Muhammad ibn Musa al-Khwarizmi's "Arithmetic", one of Leonardo's sources mentioned by Sigler, author of "Fibonacci's Liber Abaci", 2002.

- During the multiplication phase, the lattice is filled in with two-digit products of the corresponding digits labeling each row and column: the tens digit goes in the top-left corner.
- During the addition phase, the lattice is summed on the diagonals.
- Finally, if a carry phase is necessary, the answer as shown along the left and bottom sides of the lattice is converted to normal form by carrying ten's digits as in long addition or multiplication.

Example

The pictures on the right show how to calculate 345 × 12 using lattice multiplication. As a more complicated example, consider the picture below displaying the computation of 23,958,233 multiplied by 5,830 (multiplier); the result is 139,676,498,390. Notice 23,958,233 is along the top of the lattice and 5,830 is along the right side. The products fill the lattice and the sum of those products (on the diagonal) are along the left and bottom sides. Then those sums are totaled as shown. \

```
     2   3   9   5   8   2   3   3       01
   +---+---+---+---+---+---+---+---+-    002
   |1 /|1 /|4 /|2 /|4 /|1 /|1 /|1 /|     0017
   | / | / | / | / | / | / | / | / | 5   00024
 01|/ 0|/ 5|/ 5|/ 5|/ 0|/ 0|/ 5|/ 5|     000026
   +---+---+---+---+---+---+---+---+-    0000015
   |1 /|2 /|7 /|4 /|6 /|1 /|2 /|2 /|     00000013
   | / | / | / | / | / | / | / | / | 8   000000018
 02|/ 6|/ 4|/ 2|/ 0|/ 4|/ 6|/ 4|/ 4|     0000000017
   +---+---+---+---+---+---+---+---+-    00000000013
   |0 /|0 /|2 /|1 /|2 /|0 /|0 /|0 /|     000000000009
   | / | / | / | / | / | / | / | / | 3   0000000000000
 17|/ 6|/ 9|/ 7|/ 5|/ 4|/ 6|/ 9|/ 9|     -------------
   +---+---+---+---+---+---+---+---+-     139676498390
   |0 /|0 /|0 /|0 /|0 /|0 /|0 /|0 /|
   | / | / | / | / | / | / | / | / | 0
 24|/ 0|/ 0|/ 0|/ 0|/ 0|/ 0|/ 0|/ 0|     = 139,676,498,390
   +---+---+---+---+---+---+---+---+-
     26  15  13  18  17  13  09  00
```

Russian peasant multiplication

The binary method is also known as peasant multiplication, because it has been widely used by people who are classified as peasants and thus have not memorized the multiplication tables required for long multiplication.[failed verification] The algorithm was in use in ancient Egypt. Its main advantages are that it can be taught quickly, requires no memorization, and can be performed using tokens, such as poker chips, if paper and pencil aren't available. The disadvantage is that it takes more steps than long multiplication, so it can be unwieldy for large numbers.

Description

On paper, write down in one column the numbers you get when you repeatedly halve the multiplier, ignoring the remainder; in a column beside it repeatedly double the multiplicand. Cross out each row in which the last digit of the first number is even, and add the remaining numbers in the second column to obtain the product.

Examples

This example uses peasant multiplication to multiply 11 by 3 to arrive at a result of 33.

```
Decimal:      Binary:
11    3       1011   11
5     6       101   110
2    ~~12~~      10   ~~1100~~
1    24       1   11000
     ——       ——————
     33       100001
```

Describing the steps explicitly:

- 11 and 3 are written at the top
- 11 is halved (5.5) and 3 is doubled (6). The fractional portion is discarded (5.5 becomes 5).
- 5 is halved (2.5) and 6 is doubled (12). The fractional portion is discarded (2.5 becomes 2). The figure in the left column (2) is even, so the figure in the right column (12) is discarded.
- 2 is halved (1) and 12 is doubled (24).
- All not-scratched-out values are summed: 3 + 6 + 24 = 33.

The method works because multiplication is distributive, so:

$$\begin{aligned}3\times 11 &= 3\times(1\times 2^0+1\times 2^1+0\times 2^2+1\times 2^3)\\ &= 3\times(1+2+8)\\ &= 3+6+24\\ &= 33.\end{aligned}$$

A more complicated example, using the figures from the earlier examples (23,958,233 and 5,830):

```
Decimal:                 Binary:
5830   23958233          1011011000110  1011011011001001011011001
2915   47916466          101101100011  10110110110010010110110010
1457   95832932          10110110001  101101101100100101101100100
728   191665864          1011011000  1011011011001001011011001000
364   383331728          101101100  10110110110010010110110010000
182   766663456          10110110  101101101100100101101100100000
91  1533326912           1011011  1011011011001001011011001000000
45  3066653824           101101  10110110110010010110110010000000
22  6133307648           10110  101101101100100101101100100000000
11 12266615296           1011  1011011011001001011011001000000000
5  24533230592           101  10110110110010010110110010000000000
2  49066461184           10  101101101100100101101100100000000000
1  98132922368           1  1011011011001001011011001000000000000
   -----------              -------------------------------------
                            1022143253354344244353353243222210110 (before carry)
   139676498390            1000001000010101011110001110011101011 0
```

Quarter square multiplication

Two quantities can be multiplied using quarter squares by employing the following identity involving the floor function that some sources attribute to Babylonian mathematics (2000–1600 BC).

$$\left\lfloor \frac{(x+y)^2}{4} \right\rfloor - \left\lfloor \frac{(x-y)^2}{4} \right\rfloor = \frac{1}{4}\left(\left(x^2+2xy+y^2\right)-\left(x^2-2xy+y^2\right)\right) = \frac{1}{4}(4xy) = xy.$$

If one of x+y and x"y is odd, the other is odd too, thus their squares are 1 mod 4, then taking floor reduces both by a quarter; the subtraction then cancels the quarters out, and discarding the remainders does not introduce any difference comparing with the same expression without the floor functions. Below is a lookup table of quarter squares with the remainder discarded for the digits 0 through 18; this allows for the multiplication of numbers up to 9×9.\

n	0	1	2	3	4	5	6	7	8	9	10	11	12	13	14	15	16	17	18
$\lfloor n^2/4 \rfloor$	0	0	1	2	4	6	9	12	16	20	25	30	36	42	49	56	64	72	81

If, for example, you wanted to multiply 9 by 3, you observe that the sum and difference are 12 and 6 respectively. Looking both those values up on the table yields 36 and 9, the difference of which is 27, which is the product of 9 and 3.

Antoine Voisin published a table of quarter squares from 1 to 1000 in 1817 as an aid in multiplication. A larger table of quarter squares from 1 to 100000 was published by Samuel Laundy in 1856, and a table from 1 to 200000 by Joseph Blater in 1888.

Quarter square multipliers were used in analog computers to form an analog signal that was the product of two analog input signals. In this application, the sum and difference of two input voltages are formed using operational amplifiers. The square of each of these is approximated using piecewise linear circuits. Finally the difference of the two squares is formed and scaled by a factor of one fourth using yet another operational amplifier.

In 1980, Everett L. Johnson proposed using the quarter square method in a digital multiplier. To form the product of two 8-bit integers, for example, the digital device forms the sum and difference, looks both quantities up in a table of squares, takes the difference of the results, and divides by four by shifting two bits to the right. For 8-bit integers the table of quarter squares will have 2^{9}"1=511 entries (one entry for the full range 0..510 of possible sums, the differences using only the first 256 entries in range 0..255) or 2^{9}"1=511 entries (using for negative differences the technique of 2-complements and 9-bit masking, which avoids testing the sign of differences), each entry being 16-bit wide (the entry values are from $(0^2/4)=0$ to $(510^2/4)=65025$).

The quarter square multiplier technique has also benefitted 8-bit systems that do not have any support for a hardware multiplier. Charles Putney implemented this for the 6502.

Computational complexity of multiplication

Karatsuba multiplication

For systems that need to multiply numbers in the range of several thousand digits, such as computer algebra systems and bignum libraries, long multiplication is too slow. These systems may employ Karatsuba multiplication, which was discovered in 1960 (published in 1962). The heart of Karatsuba's method lies in the observation that two-digit multiplication can be done with only three rather than the four multiplications classically required. This is an example of what is now called a divide-and-conquer algorithm. Suppose we want to multiply two 2-digit base-m numbers: $x_1 m + x_2$ and $y_1 m + y_2$:

1. compute $x_1 \cdot y_1$, call the result F
2. compute $x_2 \cdot y_2$, call the result G
3. compute $(x_1 + x_2) \cdot (y_1 + y_2)$, call the result H
4. compute H ™ F ™ G, call the result K; this number is equal to $x_1 \cdot y_2 + x_2 \cdot y_1$
5. compute $F \cdot m^2 + K \cdot m + G$.

To compute these three products of base m numbers, we can employ the same trick again, effectively using recursion. Once the numbers are computed, we need to add them together (steps 4 and 5), which takes about n operations.

Karatsuba multiplication has a time complexity of $O(n^{\log_2 3})$ H™ $O(n^{1.585})$, making this method significantly faster than long multiplication. Because of the overhead of recursion, Karatsuba's multiplication is slower than long multiplication for small values of n; typical implementations therefore switch to long multiplication for small values of n.

Karatsuba's algorithm was the first known algorithm for multiplication that is asymptotically faster than long multiplication, and can thus be viewed as the starting point for the theory of fast multiplications.

Toom–Cook

Another method of multiplication is called Toom–Cook or Toom-3. The Toom–Cook method splits each number to be multiplied into multiple parts. The Toom–Cook method is one of the generalizations of the Karatsuba method. A three-way Toom–Cook can do a size-3N multiplication for the cost of five size-N multiplications. This accelerates the operation by a factor of 9/5, while the Karatsuba method accelerates it by 4/3.

Although using more and more parts can reduce the time spent on recursive multiplications further, the overhead from additions and digit management also grows. For this reason, the method of Fourier transforms is typically faster for numbers with several thousand digits, and asymptotically faster for even larger numbers.

Lower bounds

There is a trivial lower bound of Ù(n) for multiplying two n-bit numbers on a single processor; no matching algorithm (on conventional machines, that is on Turing equivalent machines) nor any sharper lower bound is known. Multiplication lies outside of $AC^0[p]$ for any prime p, meaning there is no family of constant-depth, polynomial (or even subexponential) size circuits using AND, OR, NOT, and MOD_p gates that can compute a product. This follows from a constant-depth reduction of MOD_q to multiplication. Lower bounds for multiplication are also known for some classes of branching programs.

Complex number multiplication

Complex multiplication normally involves four multiplications and two additions.

$$(a+bi)(c+di) = (ac-bd)+(bc+ad)i.$$

Or

×	*a*	*bi*
c	*ac*	*bci*
di	*adi*	*–bd*

As observed by Peter Ungar in 1963, one can reduce the number of multiplications to three, using essentially the same computation as Karatsuba's algorithm. The product (a + bi) · (c + di) can be calculated in the following way.

$k_1 = c \cdot (a + b)$

$k_2 = a \cdot (d \text{ H } c)$

$k_3 = b \cdot (c + d)$

Real part = k_1 " k_3

Imaginary part = $k_1 + k_2$.

This algorithm uses only three multiplications, rather than four, and five additions or subtractions rather than two. If a multiply is more expensive than three adds or subtracts, as when calculating by hand, then there is a gain in speed. On modern computers a multiply and an add can take about the same time so there may be no speed gain. There is a trade-off in that there may be some loss of precision when using floating point.

For fast Fourier transforms (FFTs) (or any linear transformation) the complex multiplies are by constant coefficients c + di (called twiddle factors in FFTs), in which case two of the additions (d"c and c+d) can be precomputed. Hence, only three multiplies and three adds are required. However, trading off a multiplication for an addition in this way may no longer be beneficial with modern floating-point units.

Polynomial multiplication

All the above multiplication algorithms can also be expanded to multiply polynomials. For instance the Strassen algorithm may be used for polynomial multiplication Alternatively the Kronecker substitution technique may be used to convert the problem of multiplying polynomials into a single binary multiplication.

Long multiplication methods can be generalised to allow the multiplication of algebraic formulae:

14ac - 3ab + 2 multiplied by ac - ab + 1

14ac -3ab 2

ac -ab 1

$14a^2c^2$ $-3a^2bc$ 2ac

$-14a^2bc$ $3a^2b^2$ -2ab

$14ac \quad -3ab \quad 2$

$14a^2c^2 \; -17a^2bc \quad 16ac \quad 3a^2b^2 \quad -5ab \; +2$

==

As a further example of column based multiplication, consider multiplying 23 long tons (t), 12 hundredweight (cwt) and 2 quarters (qtr) by 47. This example uses avoirdupois measures: 1 t = 20 cwt, 1 cwt = 4 qtr.

```
 t   cwt qtr
 23   12  2
        47 x
________________________
141   94  94
940  470
 29   23
________________________
1110  587 94
________________________
1110   7  2
=================
```

================= Answer: 1110 ton 7 cwt 2 qtr

First multiply the quarters by 47, the result 94 is written into the first workspace. Next, multiply cwt 12*47 = (2 + 10)*47 but don't add up the partial results (94, 470) yet. Likewise multiply 23 by 47 yielding (141, 940). The quarters column is totaled and the result placed in the second workspace (a trivial move in this case). 94 quarters is 23 cwt and 2 qtr, so place the 2 in the answer and put the 23 in the next column left. Now add up the three entries in the cwt column giving 587. This is 29 t 7 cwt, so write the 7 into the answer and the 29 in the column to the left. Now add up the tons column. There is no adjustment to make, so the result is just copied down.

The same layout and methods can be used for any traditional measurements and non-decimal currencies such as the old British £sd system.

DIVISION OF INTEGERS

Division is one of the four basic operations of arithmetic, the ways that numbers are combined to make new numbers. The other operations are addition, subtraction, and multiplication.

At an elementary level the division of two natural numbers is, among other possible interpretations, the process of calculating the number of times one number is contained within another.[:7] This number of times need not be an integer. For example, if 20 apples are divided evenly between 4 people, everyone receives 5 apples.

The division with remainder or Euclidean division of two natural numbers provides an integer quotient, which is the number of times the second number is completely contained in the first number, and a remainder, which is the part of the first number that remains, when in the course of computing the quotient, no further full chunk of the size of the second number can be allocated. For example, if 21 apples are divided between 4 people, everyone receives 5 apples again, and 1 apple remains.

For division to always yield one number rather than a quotient plus a remainder, the natural numbers must be extended to rational numbers or real numbers. In these enlarged number systems, division is the inverse operation to multiplication, that is a = c / b means a × b = c, as long as b is not zero. If b = 0, then this is a division by zero, which is not defined.[a]: 246 In the 21-apples example, everyone would receive 5 apple and a quarter of an apple, thus avoiding any leftover.

Both forms of division appear in various algebraic structures, different ways of defining mathematical structure. Those in which a Euclidean division (with remainder) is defined are called Euclidean domains and include polynomial rings in one indeterminate (which define multiplication and addition over single-variabled formulas). Those in which a division (with a single result) by all nonzero elements is defined are called fields and division rings. In a ring the elements by which division is always possible are called the units (for example, 1 and "1 in the ring of integers). Another generalization of division to algebraic structures is the quotient group, in which the result of "division" is a group rather than a number.

The simplest way of viewing division is in terms of quotition and partition: from the quotition perspective, 20 / 5 means the number of 5s that must be added to get 20. In terms of partition, 20 / 5 means the size of each of 5 parts into which a set of size 20 is divided. For example, 20 apples divide into five groups of four apples, meaning that twenty divided by five is equal to four. This is denoted as 20 / 5 = 4, or 20/5 = 4. What is being divided is called the dividend, which is divided by the divisor, and the result is called the quotient. In the example, 20 is the dividend, 5 is the divisor, and 4 is the quotient.

Unlike the other basic operations, when dividing natural numbers there is sometimes a remainder that will not go evenly into the dividend; for example, 10 / 3 leaves a remainder of 1, as 10 is not a multiple of 3. Sometimes this remainder is added to the quotient as a fractional part, so 10 / 3 is equal to 3+1/3 or 3.33..., but in the context of integer division, where numbers have no fractional part, the remainder is kept separately (or exceptionally, discarded or rounded). When the remainder is kept as a fraction, it leads to a rational number. The set of all rational numbers is created by extending the integers with all possible results of divisions of integers.

Unlike multiplication and addition, division is not commutative, meaning that a / b is not always equal to b / a. Division is also not, in general, associative, meaning that when dividing multiple times, the order of division can change the result. For example, (24 / 6) / 2 = 2, but 24 / (6 / 2) = 8 (where the use of parentheses indicates that the operations inside parentheses are performed before the operations outside parentheses).

If multiplication is totaling of numbers, division is the distribution of numbers. Dividing integers is opposite operation of multiplication. But the rules for division of integers are same as multiplication rules.Though, it is not always necessary that the quotient will always be an integer.

Rule 1: The quotient of two positive integers will always be positive.

Rule 2: The quotient of two negative integers will always be positive.

Rule 3: The quotient of a positive integer and a negative integer will always be negative.

Rules for Dividing Integers

Type of Numbers	Operation	Result	Example
Positive ÷ Positive	Divide	Positive (+)	10 ÷ 5 = 2
Negative ÷ Negative	Divide	Positive (+)	(-10) ÷ (-5) = 2
Positive ÷ Negative	Divide	Negative (-)	10 ÷ (-5) = -2
Negative ÷ Positive	Divide	Negative (-)	(-10) ÷ 5 = -2

Same as multiplication, divide the integers without the sign, then give the sign according to the rule as given in the table. The division of two integers with the like signs gives positive quotient and the division of two integers with unlike signs gives negative quotient.

To solve more problems on the topic, download BYJU'S – The Learning App from Google Play Store and watch interactive videos. Also, take free tests to practice for exams.

Notation

Plus and minuses. An obelus used as a variant of the minus sign in an excerpt from an official Norwegian trading statement form called «Næringsoppgave 1» for the taxation year 2010.

Division is often shown in algebra and science by placing the dividend over the divisor with a horizontal line, also called a fraction bar, between them. For example, "a divided by b" can written as:

$$\frac{a}{b}$$

which can also be read out loud as "divide a by b" or "a over b". A way to express division all on one line is to write the dividend (or numerator), then a slash, then the divisor (or denominator), as follows:

a/b

This is the usual way of specifying division in most computer programming languages, since it can easily be typed as a simple sequence of ASCII characters. (It is also the only notation used for quotient objects in abstract algebra.) Some mathematical software, such as MATLAB and GNU Octave, allows the operands to be written in the reverse order by using the backslash as the division operator:

b"a

A typographical variation halfway between these two forms uses a solidus (fraction slash), but elevates the dividend and lowers the divisor:

a/b

Any of these forms can be used to display a fraction. A fraction is a division expression where both dividend and divisor are integers (typically called the numerator and denominator), and there is no implication that the division must be evaluated further. A second way to show division is to use the division sign (÷, also known as obelus though the term has additional meanings), common in arithmetic, in this manner:

a ÷ b

This form is infrequent except in elementary arithmetic. ISO 80000-2-9.6 states it should not be used. This division sign is also used alone to represent the division operation itself, as for instance as a label on a key of a calculator. The obelus was introduced by Swiss mathematician Johann Rahn in 1659 in Teutsche Algebra.[: 211] The ÷ symbol is used to indicate subtraction in some European countries, so its use may be misunderstood.

In some non-English-speaking countries, a colon is used to denote division:

a : b

This notation was introduced by Gottfried Wilhelm Leibniz in his 1684 Acta eruditorum.[: 295] Leibniz disliked having separate symbols for ratio and division. However, in English usage the colon is restricted to expressing the related concept of ratios.

Since the 19th century, US textbooks have used b) a or $b\overline{)a}$ to denote a divided by b, especially when discussing long division. The history of this notation is not entirely clear because it evolved over time.

COMPUTING

Manual methods

Division is often introduced through the notion of "sharing out" a set of objects, for example a pile of lollies, into a number of equal portions. Distributing the objects several at a time in each round of sharing to each portion leads to the idea of 'chunking' – a form of division where one repeatedly subtracts multiples of the divisor from the dividend itself.

By allowing one to subtract more multiples than what the partial remainder allows at a given stage, more flexible methods, such as the bidirectional variant of chunking, can be developed as well.

More systematically and more efficiently, two integers can be divided with pencil and paper with the method of short division, if the divisor is small, or long division, if the divisor is larger. If the dividend has a fractional part (expressed as a decimal fraction), one can continue the procedure past the ones place as far as desired. If the divisor has a fractional part, one can restate the problem by moving the decimal to the right in both numbers until the divisor has no fraction, which can make the problem easier to solve (e.g., 10/2.5 = 100/25 = 4).

Division can be calculated with an abacus.

Logarithm tables can be used to divide two numbers, by subtracting the two numbers' logarithms, then looking up the antilogarithm of the result.

Division can be calculated with a slide rule by aligning the divisor on the C scale with the dividend on the D scale. The quotient can be found on the D scale where it is aligned with the left index on the C scale. The user is responsible, however, for mentally keeping track of the decimal point.

By computer

Modern calculators and computers compute division either by methods similar to long division, or by faster methods; see Division algorithm.

In modular arithmetic (modulo a prime number) and for real numbers, nonzero numbers have a multiplicative inverse. In these cases, a division by x may be computed as the product by the multiplicative inverse of x. This approach is often associated with the faster methods in computer arithmetic.

Division in different contexts

Euclidean division

Euclidean division is the mathematical formulation of the outcome of the usual process of division of integers. It asserts that, given two integers, a, the dividend, and b, the divisor, such that b '" 0, there are unique integers q, the quotient, and r, the remainder, such that a = bq + r and 0 d" r < |b|, where |b| denotes the absolute value of b.

Of integers

Integers are not closed under division. Apart from division by zero being undefined, the quotient is not an integer unless the dividend is an integer multiple of the divisor. For example, 26 cannot be divided by 11 to give an integer. Such a case uses one of five approaches:

1. Say that 26 cannot be divided by 11; division becomes a partial function.
2. Give an approximate answer as a floating-point number. This is the approach usually taken in numerical computation.
3. Give the answer as a fraction representing a rational number, so the result of the division of 26 by 11 is $\frac{26}{11}$ (or as a mixed number, so $\frac{26}{11} = 2\frac{4}{11}$.) Usually the resulting fraction should be simplified: the result of the division of 52 by 22 is also . This simplification may be done by factoring out the greatest common divisor.
4. Give the answer as an integer quotient and a remainder, so remainder 4. To make the distinction with the previous case, this division, with two integers as result, is sometimes called Euclidean division, because it is the basis of the Euclidean algorithm.
5. Give the integer quotient as the answer, so = 2. This is the floor function applied to case 2 or 3. It is sometimes called integer division, and denoted by "//".

Dividing integers in a computer program requires special care. Some programming languages, treat integer division as in case 5 above, so the answer is an integer. Other languages, such as MATLAB and every computer algebra system return a rational number as the answer, as in case 3 above. These languages also provide functions to get the results of the other cases, either directly or from the result of case 3.

Names and symbols used for integer division include div, /, \, and %. Definitions vary regarding integer division when the dividend or the divisor is negative: rounding may be toward zero (so called T-division) or toward " (F-division); rarer styles can occur – see Modulo operation for the details.

Divisibility rules can sometimes be used to quickly determine whether one integer divides exactly into another.

Of rational numbers

The result of dividing two rational numbers is another rational number when the divisor is not 0. The division of two rational numbers p/q and r/s can be computed as

$$\frac{p/q}{r/s} = \frac{p}{q} \times \frac{s}{r} = \frac{ps}{qr}.$$

All four quantities are integers, and only p may be 0. This definition ensures that division is the inverse operation of multiplication.

Of real numbers

Division of two real numbers results in another real number (when the divisor is nonzero). It is defined such that a/b = c if and only if a = cb and b 'TM 0.

Of complex numbers

Dividing two complex numbers (when the divisor is nonzero) results in another complex number, which is found using the conjugate of the denominator:

$$\frac{p+iq}{r+is} = \frac{(p+iq)(r-is)}{(r+is)(r-is)} = \frac{pr+qs+i(qr-ps)}{r^2+s^2} = \frac{pr+qs}{r^2+s^2} + i\frac{qr-ps}{r^2+s^2}.$$

This process of multiplying and dividing by r – is is called 'realisation' or (by analogy) rationalisation. All four quantities p, q, r, s are real numbers, and r and s may not both be 0.

Division for complex numbers expressed in polar form is simpler than the definition above:

$$\frac{pe^{iq}}{re^{is}} = \frac{pe^{iq}e^{-is}}{re^{is}e^{-is}} = \frac{p}{r}e^{i(q-s)}.$$

Again all four quantities p, q, r, s are real numbers, and r may not be 0.

Of polynomials

One can define the division operation for polynomials in one variable over a field. Then, as in the case of integers, one has a remainder. See Euclidean division of polynomials, and, for hand-written computation, polynomial long division or synthetic division.

Of matrices

One can define a division operation for matrices. The usual way to do this is to define A / B = AB"1, where B"1 denotes the inverse of B, but it is far more common to write out AB"1 explicitly to avoid confusion. An elementwise division can also be defined in terms of the Hadamard product.

Left and right division

Because matrix multiplication is not commutative, one can also define a left division or so-called backslash-division as A \ B = A"1B. For this to be well defined, $B^{''1}$ need not exist, however $A^{''1}$ does need to exist. To avoid confusion, division as defined by A / B = $AB^{''1}$ is sometimes called right division or slash-division in this context.

Note that with left and right division defined this way, A / (BC) is in general not the same as (A / B) / C, nor is (AB) \ C the same as A \ (B \ C). However, it holds that A / (BC) = (A / C) / B and (AB) \ C = B \ (A \ C).

Pseudoinverse

To avoid problems when $A^{''1}$ and/or $B^{''1}$ do not exist, division can also be defined as multiplication by the pseudoinverse. That is, A / B = AB^+ and A \ B = A^+B, where A^+ and B^+ denote the pseudoinverses of A and B.

Abstract algebra

In abstract algebra, given a magma with binary operation ″ (which could nominally be termed multiplication), left division of b by a (written a \ b) is typically defined as the solution x to the equation a ″ x = b, if this exists and is unique. Similarly, right division of b by a (written b / a) is the solution y to the equation y ″ a = b. Division in this sense does not require ″ to have any particular properties (such as commutativity, associativity, or an identity element).

"Division" in the sense of "cancellation" can be done in any magma by an element with the cancellation property. Examples include matrix algebras and quaternion algebras. A quasigroup is a structure in which division is always possible, even without an identity element and hence inverses. In an integral domain, where not every element need have an inverse, division by a cancellative element a can still be performed on elements of the form ab or ca by left or right cancellation, respectively.

If a ring is finite and every nonzero element is cancellative, then by an application of the pigeonhole principle, every nonzero element of the ring is invertible, and division by any nonzero element is possible. To learn about when algebras (in the technical sense) have a division operation, refer to the page on division algebras. In particular Bott periodicity can be used to show that any real normed division algebra must be isomorphic to either the real numbers R, the complex numbers C, the quaternions H, or the octonions O.

Calculus

The derivative of the quotient of two functions is given by the quotient rule:

$$\left(\frac{f}{g}\right)' = \frac{f'g - fg'}{g^2}.$$

Division by zero

Division of any number by zero in most mathematical systems is undefined, because zero multiplied by any finite number always results in a product of zero. Entry of such an expression into most calculators produces an error message. However, in certain higher level mathematics division by zero is possible by the zero ring and algebras such as wheels. In these algebras, the meaning of division is different from traditional definitions.

In mathematics, division by zero is division where the divisor (denominator) is zero. Such a division can be formally expressed as a/0, where a is the dividend (numerator). In ordinary arithmetic, the expression has no meaning, as there is no number that, when multiplied by 0, gives a (assuming a ‘“ 0); thus, division by zero is undefined. Since any number multiplied by zero is zero, the expression 0/0 is also undefined; when it is the form of a limit, it is an indeterminate form. Historically, one of the earliest recorded references to the mathematical impossibility of assigning a value to a/0 is contained in Anglo-Irish philosopher George Berkeley's criticism of infinitesimal calculus in 1734 in The Analyst ("ghosts of departed quantities").

There are mathematical structures in which a/0 is defined for some a such as in the Riemann sphere (a model of the extended complex plane) and the Projectively extended real line; however, such structures do not satisfy every ordinary rule of arithmetic (the field axioms).

In computing, a program error may result from an attempt to divide by zero. Depending on the programming environment and the type of number (e.g., floating point, integer) being divided by zero, it may generate positive or negative infinity by the IEEE 754 floating point standard, generate an exception, generate an error message, cause the program to terminate, result in a special not-a-number value, or crash.

Elementary arithmetic

When division is explained at the elementary arithmetic level, it is often considered as splitting a set of objects into equal parts. As an example, consider having ten cookies, and these cookies are to be distributed equally to five people at a table. Each person would receive $\frac{10}{5} = 2$ cookies. Similarly, if there are ten cookies, and only one person at the table, that person would receive $\frac{10}{1} = 10$ cookies.

So, for dividing by zero, what is the number of cookies that each person receives when 10 cookies are evenly distributed among 0 people at a table? Certain words can be pinpointed in the question to highlight the problem. The problem with this question is the "when". There is no way to distribute

10 cookies to nobody. Therefore, 10/0—at least in elementary arithmetic—is said to be either meaningless or undefined.

If there are, say, 5 cookies and 2 people, the problem is in "evenly distribute". In any integer partition of 5 things into 2 parts, either one of the parts of the partition will have more elements than the other or there will be a remainder (written as 5/2 = 2 r1). Or, the problem with 5 cookies and 2 people can be solved by cutting one cookie in half, which introduces the idea of fractions (5/2 = 2+1/2). The problem with 5 cookies and 0 people, on the other hand, cannot be solved in any way that preserves the meaning of "divides".

In elementary algebra, another way of looking at division by zero is that division can always be checked using multiplication. Considering the 10/0 example above, setting x = 10/0, if x equals ten divided by zero, then x times zero equals ten, but there is no x that, when multiplied by zero, gives ten (or any number other than zero). If, instead of x = 10/0, x = 0/0, then every x satisfies the question "what number x, multiplied by zero, gives zero?"

Early attempts

The Brâhmasphumasiddhânta of Brahmagupta (c. 598–668) is the earliest text to treat zero as a number in its own right and to define operations involving zero. The author could not explain division by zero in his texts: his definition can be easily proven to lead to algebraic absurdities. According to Brahmagupta,

A positive or negative number when divided by zero is a fraction with the zero as denominator. Zero divided by a negative or positive number is either zero or is expressed as a fraction with zero as numerator and the finite quantity as denominator. Zero divided by zero is zero.

In 830, Mahâvîra unsuccessfully tried to correct the mistake Brahmagupta made in his book Ganita Sara Samgraha: "A number remains unchanged when divided by zero."

ALGEBRA

The four basic operations – addition, subtraction, multiplication and division – as applied to whole numbers (positive integers), with some restrictions, in elementary arithmetic are used as a framework to support the extension of the realm of numbers to which they apply. For instance, to make it possible to subtract any whole number from another, the realm of numbers must be expanded to the entire set of integers in order to incorporate the negative integers. Similarly, to support division of any integer by any other, the realm of numbers must expand to the rational numbers. During this gradual expansion of the number system, care is taken to ensure that the "extended operations", when applied to the older numbers, do not produce different results. Loosely speaking, since division by zero has no meaning (is undefined) in the whole number setting, this remains true as the setting expands to the real or even complex numbers.

As the realm of numbers to which these operations can be applied expands there are also changes in how the operations are viewed. For instance, in the realm of integers, subtraction is no longer considered a basic operation since it can be replaced by addition of signed numbers. Similarly, when the realm of numbers expands to include the rational numbers, division is replaced by multiplication by certain rational numbers. In keeping with this change of viewpoint, the question, "Why can't we divide by zero?", becomes "Why can't a rational number have a zero denominator?". Answering this revised question precisely requires close examination of the definition of rational numbers.

In the modern approach to constructing the field of real numbers, the rational numbers appear as an intermediate step in the development that is founded on set theory. First, the natural numbers (including zero) are established on an axiomatic basis such as Peano's axiom system and then this is expanded to the ring of integers. The next step is to define the rational numbers keeping in mind that this must be done using only the sets and operations that have already been established, namely, addition, multiplication and the integers. Starting with the set of ordered pairs of integers, {(a, b)} with b ′″ 0, define a binary relation on this set by (a, b) C″ (c, d) if and only if ad = bc. This relation is shown to be an equivalence relation and its equivalence classes are then defined to be the rational numbers. It is in the formal proof that this relation is an equivalence relation that the requirement that the second coordinate is not zero is needed (for verifying transitivity).

The above explanation may be too abstract and technical for many purposes, but if one assumes the existence and properties of the rational numbers, as is commonly done in elementary mathematics, the "reason" that division by zero is not allowed is hidden from view. Nevertheless, a (non-rigorous) justification can be given in this setting.

It follows from the properties of the number system we are using (that is, integers, rationals, reals, etc.), if b ′″ 0 then the equation $a/b = c$ is equivalent to $a = b \times c$. Assuming that $a/0$ is a number c, then it must be that $a = 0 \times c = 0$. However, the single number c would then have to be determined by the equation $0 = 0 \times c$, but every number satisfies this equation, so we cannot assign a numerical value to $0/0$.

HIGHER MATHEMATICS

Although division by zero cannot be sensibly defined with real numbers and integers, it is possible to consistently define it, or similar operations, in other mathematical structures.

Non-standard analysis

In the hyperreal numbers and the surreal numbers, division by zero is still impossible, but division by non-zero infinitesimals is possible.

Distribution theory

In distribution theory one can extend the function $1/x$ to a distribution on the whole space of real numbers (in effect by using Cauchy principal values). It does not, however, make sense to ask for a "value" of this distribution at $x = 0$; a sophisticated answer refers to the singular support of the distribution.

Linear algebra

In matrix algebra (or linear algebra in general), one can define a pseudo-division, by setting $a/b = ab^+$, in which b^+ represents the pseudoinverse of b. It can be proven that if b^{-1} exists, then $b^+ = b^{-1}$. If b equals 0, then $b^+ = 0$.

Abstract algebra

In abstract algebra, the integers, the rational numbers, the real numbers, and the complex numbers can be abstracted to more general algebraic structures, such as a commutative ring, which is a mathematical structure where addition, subtraction, and multiplication behave as they do in the more familiar number systems, but division may not be defined. Adjoining a multiplicative inverses to a commutative ring is called localization. However, the localization of every commutative ring at zero is the trivial ring, where 0 = 1, so nontrivial commutative rings do not have inverses at zero, and thus division by zero is undefined for nontrivial commutative rings.

Nevertheless, any number system that forms a commutative ring can be extended to a seldom used structure called a wheel in which division by zero is always possible. However, the resulting mathematical structure is no longer a commutative ring, as multiplication no longer distributes over addition. Furthermore, in a wheel, division of an element by itself no longer results in the multiplicative identity element 1, and if the original system was an integral domain, the multiplication in the wheel no longer results in a cancellative semigroup.

The concepts applied to standard arithmetic are similar to those in more general algebraic structures, such as rings and fields. In a field, every nonzero element is invertible under multiplication; as above, division poses problems only when attempting to divide by zero. This is likewise true in a skew field (which for this reason is called a division ring). However, in other rings, division by nonzero elements may also pose problems. For example, the ring Z/6Z of integers mod 6. The meaning of the expression 2/2 should be the solution x of the equation 2x = 2. But in the ring Z/6Z, 2 is a zero divisor. This equation has two distinct solutions, x = 1 and x = 4, so the expression 2/2 is undefined.

In field theory, the expression a/b is only shorthand for the formal expression $ab^{\prime\prime 1}$, where $b^{\prime\prime 1}$ is the multiplicative inverse of b. Since the field axioms only guarantee the existence of such inverses for nonzero elements, this expression has no meaning when b is zero. Modern texts, that define fields as a special type of ring, include the axiom 0 '" 1 for fields (or its equivalent) so that the zero ring is excluded from being a field. In the zero ring, division by zero is possible, which shows that the other field axioms are not sufficient to exclude division by zero in a field.

Computer arithmetic

The IEEE floating-point standard, supported by almost all modern floating-point units, specifies that every floating point arithmetic operation, including division by zero, has a well-defined result. The standard supports signed zero, as well as infinity and NaN (not a number). There are two zeroes: +0 (positive zero) and "0 (negative zero) and this removes any ambiguity when dividing. In IEEE 754 arithmetic, a ÷ +0 is positive infinity when a is positive, negative infinity when a is negative, and NaN when a = ±0. The infinity signs change when dividing by "0 instead.

The justification for this definition is to preserve the sign of the result in case of arithmetic underflow. For example, in the single-precision computation 1/(x/2), where x = $\pm 2^{\prime\prime 149}$, the computation x/2 underflows and produces ±0 with sign matching x, and the result will be ±" with sign matching x. The sign will match that of the exact result $\pm 2^{150}$, but the magnitude of the exact result is too large to represent, so infinity is used to indicate overflow.

Integer division by zero is usually handled differently from floating point since there is no integer representation for the result. Some processors generate an exception when an attempt is made to divide an integer by zero, although others will simply continue and generate an incorrect result for the division. The result depends on how division is implemented, and can either be zero, or sometimes the largest possible integer.

Because of the improper algebraic results of assigning any value to division by zero, many computer programming languages (including those used by calculators) explicitly forbid the execution of the operation and may prematurely halt a program that attempts it, sometimes reporting a "Divide by zero" error. In these cases, if some special behavior is desired for division by zero, the condition must be explicitly tested (for example, using an if statement). Some programs (especially those that use fixed-point arithmetic where no dedicated floating-point hardware is available) will use behavior similar to the IEEE standard, using large positive and negative numbers to approximate infinities. In some programming languages, an attempt to divide by zero results in undefined behavior. The graphical programming language Scratch 2.0 and 3.0 used in many schools returns Infinity or "Infinity depending on the sign of the dividend.

In two's complement arithmetic, attempts to divide the smallest signed integer by "1 are attended by similar problems, and are handled with the same range of solutions, from explicit error conditions to undefined behavior.

Most calculators will either return an error or state that 1/0 is undefined; however, some TI and HP graphing calculators will evaluate $(1/0)^2$ to ".

Microsoft Math and Mathematica return ComplexInfinity for 1/0. Maple and SageMath return an error message for 1/0, and infinity for 1/0.0 (0.0 tells these systems to use floating point arithmetic instead of algebraic arithmetic).

Some modern calculators allow division by zero in special cases, where it will be useful to students and, presumably, understood in context by mathematicians. Some calculators, the online Desmos calculator is one example, allow arctangent(1/0). Students are often taught that the inverse cotangent function, arccotangent, should be calculated by taking the arctangent of the reciprocal, and so a calculator may allow arctangent(1/0), giving the output $\pi/2$, which is the correct value of arccotangent 0. The mathematical justification is that the limit as x goes to zero of arctangent $1/x$ is $\pi/2$.

Bibliography

- "In equall Parallelipipedons the bases are reciprokall to their altitudes". OED "Reciprocal" §3a. Sir Henry Billingsley translation of Elements XI, 34.
- Anthony, Dr. "Proof that INT(1/x)dx = lnx". Ask Dr. Math. Drexel University. Retrieved 22 March 2013.
- Mitchell, Douglas W., "A nonlinear random number generator with known, long cycle length," Cryptologia 17, January 1993, 55–62.
- Blake, A. G. (1887). Arithmetic. Dublin, Ireland: Alexander Thom & Company.
- Derbyshire, John (2004). Prime Obsession: Bernhard Riemann and the Greatest Unsolved Problem in Mathematics. New York City: Penguin Books. ISBN 978-0-452-28525-5.
- http://www.mathwords.com/a/associative_operation.htm Archived 2018-10-28 at the Wayback Machine Retrieved October 23, 2018
- George Mark Bergman: Order of arithmetic operations Archived 2017-03-05 at the Wayback Machine
- Education Place: The Order of Operations Archived 2017-06-08 at the Wayback Machine
- Cajori, Florian (1929). A History of Mathematical Notations. Open Court Pub. Co.
- Symbol and Notation, (Weidenfeld & Nicolson, London: 1972), 81ff.
- "Egyptian Mathematical Papyri – Mathematicians of the African Diaspora". Math.buffalo.edu. Archived from the original on 2015-04-07. Retrieved 2012-01-30.
- "Historia Matematica Mailing List Archive: Re: [HM] The Zero Story: a question". Sunsite.utk.edu. 1999-04-26. Archived from the original on 2012-01-12. Retrieved 2012-01-30.
- Sánchez, George I. (1961). Arithmetic in Maya. Austin, Texas: self published.

- Staszkow, Ronald; Robert Bradshaw (2004). The Mathematical Palette (3rd ed.). Brooks Cole. p. 41. ISBN 0-534-40365-4.
- Smith, David Eugene (1958). History of Modern Mathematics. Dover Publications. p. 259. ISBN 0-486-20429-4.
- "Classical Greek culture (article)". Khan Academy. Retrieved 2022-05-04.
- Selin, Helaine, ed. (2000). Mathematics across cultures: the history of non-Western mathematics. Kluwer Academic Publishers. p. 451. ISBN 0-7923-6481-3.
- Bernard Frischer (1984). "Horace and the Monuments: A New Interpretation of the Archytas Ode". In D.R. Shackleton Bailey (ed.). Harvard Studies in Classical Philology. Harvard University Press. p. 83. ISBN 0-674-37935-7.
- Eduard Heine, "Die Elemente der Functionenlehre", [Crelle's] Journal für die reine und angewandte Mathematik, [1] 74 (1872): 172–188.
- Georg Cantor, "Ueber unendliche, lineare Punktmannichfaltigkeiten", pt. 5, Mathematische Annalen, 21, 4 (1883 12): 545–591.
- Richard Dedekind, Stetigkeit & irrationale Zahlen (Braunschweig: Friedrich Vieweg & Sohn, 1872). Subsequently published in: ———, Gesammelte mathematische Werke, ed. Robert Fricke, Emmy Noether & Öystein Ore (Braunschweig: Friedrich Vieweg & Sohn, 1932), vol. 3, pp. 315–334.
- L. Kronecker, "Ueber den Zahlbegriff", [Crelle's] Journal für die reine und angewandte Mathematik, [1] 101 (1887): 337–355.
- Leonhard Euler, "Conjectura circa naturam aeris, pro explicandis phaenomenis in atmosphaera observatis", Acta Academiae Scientiarum Imperialis Petropolitanae, 1779, 1 (1779): 162–187.
- Ramus, "Determinanternes Anvendelse til at bes temme Loven for de convergerende Bröker", in: Det Kongelige Danske Videnskabernes Selskabs naturvidenskabelige og mathematiske Afhandlinger (Kjoebenhavn: 1855), p. 106.
- Eduard Heine, "Einige Eigenschaften der Laméschen Funktionen", [Crelle's] Journal für die reine und angewandte Mathematik, [1] 56 (Jan. 1859): 87–99 at 97.
- Siegmund Günther, Darstellung der Näherungswerthe von Kettenbrüchen in independenter Form (Erlangen: Eduard Besold, 1873); ———, "Kettenbruchdeterminanten", in: Lehrbuch der Determinanten-Theorie: Für Studirende (Erlangen: Eduard Besold, 1875), c. 6, pp. 156–186.
- Bogomolny, A. "What's a number?". Interactive Mathematics Miscellany and Puzzles. Archived from the original on 23 September 2010. Retrieved 11 July 2010.

- "Eine Menge, ist die Zusammenfassung bestimmter, wohlunterschiedener Objekte unserer Anschauung oder unseres Denkens – welche Elemente der Menge genannt werden – zu einem Ganzen." [https://web.archive.org/web/20110610133240/http://brinkmann-du.de/mathe/fos/fos01_03.htm Archived 2011-06-10 at the Wayback Machine]
- "natural number", Merriam-Webster.com, Merriam-Webster, archived from the original on 13 December 2019, retrieved 4 October 2014
- Suppes, Patrick (1972). Axiomatic Set Theory. Courier Dover Publications. p. 1. ISBN 0-486-61630-4.
- Weisstein, Eric W. "Repeating Decimal". mathworld.wolfram.com. Retrieved 2020-07-23
- Thomas Sonnabend (2010). Mathematics for Teachers: An Interactive Approach for Grades K–8. Brooks/Cole, Cengage Learning (Charles Van Wagner). p. 126. ISBN 978-0-495-56166-8.
- Smith, David Eugene (1925). History Of Mathematics Vol II. Ginn And Company.
- Jakub Czajko (July 2004) "On Cantorian spacetime over number systems with division by zero", Chaos, Solitons and Fractals, volume 21, number 2, pages 261–271.
- Ben Goldacre (2006-12-07). "Maths Professor Divides By Zero, Says BBC".
- To Continue with Continuity Metaphysica 6, pp. 91–109, a philosophy paper from 2005, reintroduced the (ancient Indian) idea of an applicable whole number equal to 1/0, in a more modern (Cantorian) style.
- Kojima, Takashi (2012-07-09). Advanced Abacus: Theory and Practice. Tuttle Publishing. ISBN 978-1-4629-0365-8.
- "In equall Parallelipipedons the bases are reciprokall to their altitudes". OED "Reciprocal" §3a. Sir Henry Billingsley translation of Elements XI, 34.
- Anthony, Dr. "Proof that INT(1/x)dx = lnx". Ask Dr. Math. Drexel University. Retrieved 22 March 2013.
- Mitchell, Douglas W., "A nonlinear random number generator with known, long cycle length," Cryptologia 17, January 1993, 55–62.
- http://mathworld.wolfram.com/DivisionbyZero.html Archived 2018-10-23 at the Wayback Machine Retrieved October 23, 2018
- Jesper Carlström. "On Division by Zero" Archived 2019-08-17 at the Wayback Machine Retrieved October 23, 2018
- Dunham, William (2015), The Calculus Gallery: Masterpieces from Newton to Lebesgue, Princeton University Press, p. 127, ISBN 978-1-4008-6679-3, archived from the original on 2015-05-14, retrieved 2015-

02-17, Cantor found a remarkable shortcut to reach Liouville's conclusion with a fraction of the work

- Hurwitz, Adolf (1893). "Beweis der Transendenz der Zahl e". Mathematische Annalen (43): 134–35.
- Gordan, Paul (1893). "Transcendenz von e und p". Mathematische Annalen. 43 (2–3): 222–224. doi:10.1007/bf01443647. S2CID 123203471.
- "Lecture #1" (PDF). 18.095 Lecture Series in Mathematics. 2015-01-05.
- Moschovakis, Yiannis N. (1980), "Descriptive set theory", Studies in Logic and the Foundations of Mathematics, Amsterdam; New York: North-Holland Publishing Co., vol. 100, pp. xii, 637, ISBN 978-0-444-85305-9, chapter V.
- Bishop, Errett; Bridges, Douglas (1985), Constructive analysis, Grundlehren der Mathematischen Wissenschaften [Fundamental Principles of Mathematical Sciences], vol. 279, Berlin, New York: Springer-Verlag, ISBN 978-3-540-15066-4, chapter 2.
- Wheeler, John Archibald (1986). "Hermann Weyl and the Unity of Knowledge: In the linkage of four mysteries—the "how come" of existence, time, the mathematical continuum, and the discontinuous yes-or-no of quantum physics—may lie the key to deep new insight". American Scientist. 74 (4): 366–75. Bibcode:1986AmSci..74..366W. JSTOR 27854250.
- Bengtsson, Ingemar (2017). "The Number Behind the Simplest SIC-POVM". Foundations of Physics. 47 (8): 1031–41. arXiv:1611.09087. Bibcode:2017FoPh...47.1031B. doi:10.1007/s10701-017-0078-3. S2CID 118954904.
- Hardy, G., Littlewood J. E., Pólya, G. (1999). Inequalities. Cambridge Mathematical Library, Cambridge University Press. ISBN 0-521-05206-8.
- Beckenbach, E. F., Bellman, R. (1975). An Introduction to Inequalities. Random House Inc. ISBN 0-394-01559-2.
- Drachman, Byron C., Cloud, Michael J. (1998). Inequalities: With Applications to Engineering. Springer-Verlag. ISBN 0-387-98404-6.
- Grinshpan, A. Z. (2005), "General inequalities, consequences, and applications", Advances in Applied Mathematics, 34 (1): 71–100, doi:10.1016/j.aam.2004.05.001
- Arthur Lohwater (1982). "Introduction to Inequalities". Online e-book in PDF format.
- Harold Shapiro (2005). "Mathematical Problem Solving". The Old Problem Seminar. Kungliga Tekniska högskolan.
- "3rd USAMO". Archived from the original on 2008-02-03.

Index